Women Driven Mobility: Rethinking the Way the World Moves

KATELYN SHELBY DAVIS AND KRISTIN SHAW

Warrendale, Pennsylvania, USA

400 Commonwealth Drive
Warrendale, PA 15096-0001 USA
E-mail: CustomerService@sae.org
Phone: 877-606-7323 (inside USA and Canada)
724-776-4970 (outside USA)
FAX: 724-776-0790

Copyright © 2021 SAE International. All rights reserved.

No part of this publication may be reproduced, stored in a retrieval system, or transmitted, in any form or by any means, electronic, mechanical, photocopying, recording, or otherwise, without the prior written permission of SAE International. For permission and licensing requests, contact SAE Permissions, 400 Commonwealth Drive, Warrendale, PA 15096-0001 USA; e-mail: copyright@sae.org; phone: 724-772-4028.

Library of Congress Catalog Number 2021948644
http://dx.doi.org/10.4271/9781468603095

Information contained in this work has been obtained by SAE International from sources believed to be reliable. However, neither SAE International nor its authors guarantee the accuracy or completeness of any information published herein and neither SAE International nor its authors shall be responsible for any errors, omissions, or damages arising out of use of this information. This work is published with the understanding that SAE International and its authors are supplying information but are not attempting to render engineering or other professional services. If such services are required, the assistance of an appropriate professional should be sought.

ISBN-Print 978-1-4686-0308-8
ISBN-PDF 978-1-4686-0309-5
ISBN-ePub 978-1-4686-0310-1

To purchase bulk quantities, please contact: SAE Customer Service

E-mail: CustomerService@sae.org
Phone: 877-606-7323 (inside USA and Canada)
724-776-4970 (outside USA)
Fax: 724-776-0790

Visit the SAE International Bookstore at books.sae.org

Chief Growth Officer
Frank Menchaca

Publisher
Sherry Dickinson Nigam

Development Editor
Publishers Solutions, LCC
Albany, NY

Director of Content Management
Kelli Zilko

Production and Manufacturing Associate
Erin Mendicino

Women Driven Mobility: Rethinking the Way the World Moves

To the little girls gifted an Easy Bake when they wanted Hot Wheels;
To the loud, bossy, and improper young ladies talking out of turn;
To the womxn silently pulling their own seat up to the table in the boardroom;
To those who fought the fight before we called it one;
To the boys, men, and allies who stood with us and stepped aside to make room;
To the forgotten, the marginalized, and silenced;
To her, to him, to them;
And especially to those saying "no," who look the other way when greatness glows and make the choice to refuse to accept that our modern history has too many holes:
With love, for you, for our future.

Note from the Authors

Everyone sees the world differently, through their eyes, their experiences. Women have different experiences then men—specifically in how they have been treated, how they will take risks when moving around our built environments, and their perceptions of safety. In the male-dominated planning practice, there are different priorities, concerns, and solutions. Even more varied are the perspectives of people of color who have been historically left out of planning processes. When I look at the built environment, I am concerned with where strollers, playful and active children, and seniors could stay safe against roadways and speeding cars. Inclusion means bringing all voices to the table—this needs to include race, gender, identity, age, and experience. We can create more accessible and inclusive systems and spaces that add value and mitigate damages to communities by planning and designing with all perceptions and lived experiences considered.

—Jennifer Pangborn, AICP, PTP,
Senior Supervising Planner,
Assistant Vice President, WSP USA

This book is not the answer, the word, or the solution. We only hope it is a step in the progress toward an equitable, beautiful, and peaceful world for all of us.

As two white, straight, American-born women, we wrote this book with love, in learning, and through our limited perspectives with the help of many others in the community providing guidance and lighting our way.

This work is a testimonial to the work done when the office is empty, the coffee is cold, the kids are asleep, and after hearing "absolutely not" for the tenth time. We wrote it to turn up the volume on the many stories that go untold. We wrote it to honor the shoulders we stand on and to shine a light on the footsteps that will follow.

Contents

Note from the Authors — vii
Foreword — xix

Introduction — 1

CHAPTER 1
Awareness and Community Advocacy — 7

Census Transportation Planning Products Program — 9
    Background — 9
    Goals and Objectives — 10
    Approach and Challenges — 10
    Results — 10
    About the Team — 11

Closing EV Workforce and Ownership Gender Gaps — 12
    Background — 12
    Goals and Objectives — 12
    Approach and Challenges — 13
    Results — 13
    About the Team — 13

Michigan Long-Range Transportation Plan Public Engagement Partnership for Inclusive Participation — 15
    Background — 15
    Goals and Objectives — 17
    Approach and Challenges — 18
    Results — 19
    About the Team — 19

CHAPTER 2

Design and Engineering 21

Creating Accessible AV Transportation for Those with Disabilities or Impairments and for Seniors   **23**
- Background   23
- Goals and Objectives   24
- Approach and Challenges   24
- Results   25
- Next Steps   25
- About the Team   25

Sustainable Material behind Fashionable, Handcrafted Interiors   **26**
- Background   26
- Goals and Objectives   26
- Approach and Challenges   27
- Results   27
- About the Team   27

A Smartphone-Human Hybrid Intelligence System to Measure and Shape Happiness, 2020   **28**
- Background   28
- Goals and Objectives   29
- Approach and Challenges   29
- Results   32
- About the Team   34
- References   34

From Revolutions per Minute to Experiences per Mile   **35**
- Background   35
- Goals and Objectives   36
- Approach and Challenges   36
- Results   37
- About the Team   38

CHAPTER 3

Funding 39

An Economic Development Approach to Funding Mobility Projects, 2018–2020   **41**
- Background   41
- Goals and Objectives   42

Approach and Challenges	42
Results	43
Next Steps	43
About the Team	44

Regional Transit Authority of Southeast Michigan Ballot Initiative — 44
Background	45
Goals and Objectives	45
Approach and Challenges	47
Results	48
About the Team	48

Creating Safe Pedestrian and Bike Routes through Tribal Land, Karuk Tribe — 49
Background	49
Goals and Objectives	50
Approach and Challenges	51
Results and Next Steps	52
About the Team	52

Tennessee Corridor Fast-Charging Network — 54
Background	54
Goals and Objectives	54
Approach and Challenges	55
Results	56
About the Team	57

CHAPTER 4
Infrastructure — 59

Infrastructure Connections Making a Difference in Our Communities: I-75 Modernization Project — 60
Background	61
Goals and Objectives	62
Approach and Challenges	62
Results	63
About the Team	64

MOVE: Mobility Optimization through Vision and Excellence — 66
Background	66
Goals and Objectives	66
Approach and Challenges	68

xii   Contents

Results	68
About the Team	69
Open Access Electric Vehicle Charging	**69**
Background	69
Goals and Objectives	70
Approach and Challenges	70
Results	71
About the Team	72

CHAPTER 5

Marketing and Communications   75

Bringing Back Bronco, 2021	**76**
Background	77
Goals and Objectives	79
Approach and Challenges	79
Results	80
About the Team	81
"No Way, Norway" Superbowl LV Campaign, 2021	**81**
Background	81
Goals and Objectives	82
Approach and Challenges	82
Results	83
About the Team	84
Multicultural Marketing "Built Phenomenally," 2020	**84**
Background	85
Goals and Objectives	85
Approach and Challenges	85
Results	87
About the Team	87

CHAPTER 6

Mobility on Demand   89

Limited Access Connections	**91**
Background	91
Goals and Objectives	91
Approach and Challenges	92

Results	92
About the Team	93
Bedrock's Alternative Mode of Transportation Campaign	**93**
Background	93
Goals and Objectives	94
Approach and Challenges	94
Results	95
About the Team	96
FlexLA: FASTLinkDTLA, 2018	**96**
Background	96
Goals and Objectives	98
Approach and Challenges	99
Results	100
About the Team	101

CHAPTER 7
Placemaking 103

Building a Vibrant Urban Innovation District	**105**
Background	106
Goals and Objectives	106
Approach and Challenges	107
Next Steps	108
About the Team	109
Development without Displacement: Creating 20-Minute Neighborhoods	**110**
Background	110
Goals and Objectives	111
Approach and Challenges	112
Results	113
About the Team	114
Designing Community-First Transit, 2010–2018	**115**
Background	115
Goals and Objectives	116
Approach and Challenges	116
Results	117
About the Team	117

CHAPTER 8
Policy and Legislation — 119

Commission on the Future of Mobility — 121
    Background — 121
    Goals and Objectives — 122
    Approach and Challenges — 123
    About the Team — 126

Remixing Innovation for Mobility Justice, 2020 — 126
    Background — 127
    Goals and Objectives — 128
    Approach and Challenges — 128
    Results — 131
    About the Team — 134

CHAPTER 9
Sustainability and Climate Resilience — 137

Achieving Sustainability and Equity through Electrified Micromobility — 139
    Background — 140
    Goals and Objectives — 140
    Approach and Challenges — 142
    Results — 142
    About the Team — 143

Shaping City-Scale Urban Sustainability and Quality of Life with LEED for Cities — 143
    Background — 143
    Goals and Objectives — 144
    Approach and Challenges — 144
    Results — 146
    About the Team — 147
    Additional Commentary — 147

A Sustainable and Autonomous Future in Hawaii — 148
    Background — 148
    Goals and Objectives — 149
    Approach and Challenges — 150

Results	150
About the Team	151
Climate Preparedness and Resilience Exercise Series	**153**
Background	153
Goals and Objectives	154
Approach and Challenges	155
Results	156
Additional Commentary	157
About the Team	159

CHAPTER 10
Talent and Education — 161

Championing Diversity, Equity, and Inclusion across the Industry	**162**
Background	163
Goals and Objectives	163
Approach and Challenges	164
Next Steps	164
About the Team	165
Designing a Mobile Future	**165**
Background	166
Goals and Objectives	166
Approach and Challenges	167
Results	170
About the Team	171
Preparing the Workforce for the Next Generation of Mobility	**172**
Background	172
Goals and Objectives	172
Approach and Challenges	173
Results	174
About the Team	174
Mobility Hubs of the Future	**175**
Background	175
Goals and Objectives	175
Approach and Challenges	176
Results	178
About the Team	178

CHAPTER 11

Technology Innovation — 183

Mapping the Skies for Intelligent Air Mobility — 184
    Background — 184
    Goals and Objectives — 186
    Approach and Challenges — 186
    Results — 187
    About the Team — 187

Autonomous Bus Pilot Deployment — 188
    Background — 188
    Goals and Objectives — 189
    Approach and Challenges — 190
    Results — 190
    About the Team — 191
    Additional Commentary — 193

Autonomous Trucking, 2018 — 194
    Background — 194
    Goals and Objectives — 194
    Approach and Challenges — 195
    Results — 196
    About the Team — 196
    Additional Commentary — 196

CHAPTER 12

Pivoting for the Pandemic — 197

Transportation Assistance for Tribal Communities in Remote Communities during COVID-19 — 199
    Background — 199
    Goals and Objectives — 199
    Approach and Challenges — 200
    Next Steps — 200
    About the Team — 200

Pivoting a Mobility Venture Incubator to Serve the Community — 201
    Background — 201
    Goals and Objectives — 202
    Approach and Challenges — 202

Results and Next Steps	203
About the Team	203
City of Detroit's COVID-19 Transportation Response	**204**
Background	204
Goals and Objectives	204
Approach and Challenges	204
Results	205
About the Team	206
Afterword	**207**
About the Authors	**217**
Special Thanks	**221**
About the Cover Artist	**225**
Index	**227**

Foreword

Governor Gretchen Whitmer
State of Michigan

The history and legacy of Michigan and the automobile are inextricably linked—we are the state that not only put the world on wheels but also made transportation more accessible by laying the first mile of concrete highway, more affordable with the mass production of the Model T, and safer, installing the first three-color traffic light—all in the early part of the twentieth century. As the center of the automotive industry, Michigan also drove the explosive growth of the country's middle class and implemented safer working conditions across and beyond manufacturing.

Courtesy of the Office of Governor Gretchen Whitmer.

Fast forward 100 years, give or take, and we are continuing to build that legacy, harnessing it once again as a driving force in the next generation of transportation. Instead of concrete highways, we are working to deploy the world's first fully connected autonomous vehicle corridor. Instead of the Model T, we have next-generation, fully electric vehicles rolling off production lines in places like Factory Zero in Hamtramck. And instead of traffic lights, we are not only fixing the damn roads, we are staying ahead of the curve and installing smart infrastructure that will reduce crashes and create safer roads in the coming years.

Our signature industry has a new and passionate focus on sustainability. We are putting environmentally conscious transportation systems at the forefront of policy and planning in ways we have never done before. We are proving that it is possible to achieve key climate goals and maintain a strong automotive industry by supporting the transition from internal combustion engine vehicles to electric vehicles and expanding access to charging infrastructure.

But, in all of this, we cannot lose sight of why these groundbreaking innovations and revolutions in the policies and practices of moving people, goods, and services on air, land, or sea really matter, and that is improving lives and advancing our goals of achieving a more equitable society.

In Michigan, we believe we can offer a blueprint for the rest of the nation on how to optimize mobility and electrification to improve transportation options for citizens who may not be looking for a self-driving car but who do need an affordable, reliable way to go to get to work or school, to make it to their doctor's office for appointments, or to the bank, the grocery store, or countless other places that those with reliable transportation take for granted.

If it works in Michigan—with our diverse mix of people and cultures, our four-season weather, and geography that spans dense urban environment to vast rural regions—we are confident those solutions can, and will, translate to many other markets across the United States and beyond.

That is why, while we are doubling down on our efforts to be a leader in mobility and electrification, we are also working collaboratively and cooperatively on the regional, national, and global levels to develop, test, pilot, and deploy the technologies that will transform transportation this century. That is the spirit and the mentality you will find throughout these chapters—a coming together across the public sector, private sector, philanthropic sector, and university and talent development sectors to find new solutions to challenges in transportation systems that continue to exist today.

And while the technology is often inspiring and exciting in its own right, it is just as important that we keep a clear line of sight on how it can be used to make life better for people.

Too often when we talk about mobility, the conversation goes straight to self-driving cars and autonomous technology of the future. But what we need to be talking about is how the technology and innovation happening in the mobility sector can impact and improve the lives of citizens right now.

Mobility solutions, including driverless cars, offer the opportunity to grant us all precious autonomy—making it easier for veterans, for those who are not fully able, and for our senior citizens to completely participate in society.

This is not about solving problems we anticipate tomorrow. Applied autonomy can solve real accessibility challenges facing society today.

In Michigan, we have supported the launch of a self-driving paratransit shuttle to transport senior citizens and underserved communities to and from appointments at a Detroit hospital. We provided a grant to pilot a connected, electric, and autonomous disinfecting robot ideal for electrostatically spraying down large spaces and surfaces with FDA-approved disinfection products in an effort to combat the COVID-19 virus. It has since been deployed at major international airport hubs in the United States. We are also thinking about how to leverage autonomous vehicles to bring essential goods and services to people who don't have access to transportation.

But the fact remains, all the technology in the world can only do so much if it remains unaffordable, inaccessible, or fails to address the real-world challenges people face to accessing transit here in Michigan and across the nation.

We can address that reality head-on by acknowledging the people closest to the problem are most likely also the ones closest to the solution because they know what actual end-use success looks like. We need to look around the table to make sure that the people living and confronting challenges are present when it comes to considering the problem to be solved, as well as the decision-making of how to solve it. When we do, we can identify better, more robust, more sustainable answers to pressing questions—including how we can implement a mobility reality that meets the needs of all people.

The women featured across these pages do just that. They are blazing new trails within the mobility and electrification sector—and planting a clear flag in the ground and throughout the industry that this century, women will be working to lead the charge across industries such as automotive and tech and across the public sector.

That's exciting, because here in Michigan, we know a thing or two about the unique challenges women continue to face today—from boardrooms to elected office—but we also know how much impact we are having. Today, Mary Barra is at the helm of GM helping to shape and drive a strategy that secures the company's future by investing in electrification and mobility strategies. Industry leaders like Jessica Robinson are accelerating workforce development initiatives critical to the continued growth of the industry. Policymakers, including U.S. Energy Secretary Jennifer Granholm, U.S. Senator Debbie Stabenow, and Michigan Senator Mallory McMorrow are championing initiatives to ensure our laws and policies keep up with the ecosystem in which they operate.

As the second woman elected Governor of Michigan, I was proud to be at the top of a ticket that made history when women swept all three major statewide offices, and I am proud to serve and lead our state alongside Attorney General Dana Nessel and Secretary of State Jocelyn Benson.

Together, we have become adept at taking sentiments meant to undermine, trivialize, or dismiss and instead use them to motivate, to inspire, and to drive the change we want to see in the world. We are ALL "That Woman from Michigan" at some point, and the body of work across the nation demonstrated in this book shows the consequence and results that happen when women take their seat at the table.

The women in this book are not only making new spaces in boardrooms, the startup ecosystem, universities, and every level of government—they are approaching mobility and electrification from a position of solving real-world problems that improve and enrich access to society for people from all walks of life.

It is more important than ever that they do. It is shaping up to be a landmark decade for mobility and electrification—a decade in which software will come to represent more than 50% of the value of the vehicle, electric vehicle sales are expected to pass internal combustion engine sales, and more than 50% of vehicles built will be partially autonomous.

As technology continues to evolve and transportation options grow, mobility leaders across the nation must continue to collaborate and bring the industry together rather than compete across a fragmented landscape. I am proud of the work we are doing in Michigan to lead that innovation through a public/private partnership approach as we create the environment for mobility and EV solutions to develop, grow, and be exported around the world. And I am excited by the work showcased throughout this book—to see the great minds of women across the nation driving this century's great revolution in transportation. Let's keep moving forward, together.

Introduction

One hundred years ago, the world saw the biggest disruption to date in the transportation industry—the birth of the automobile. Now, more than a century later, we are once again experiencing the magic that comes with reimagining mobility and the way that it is able to transform society. With the help of leading-edge technology, the industry is experiencing multiple disruptions, and industry leaders are rethinking the possibilities and principles of transportation.

We are on the precipice of a new era fueled by rapid technological innovations, unlimited digital connectivity, and a need for environmentally responsible and sustainable options. It is not too difficult to imagine a future in which autonomous shuttles are used for morning commutes by navigating on a specialized infrastructure designed for connected and autonomous vehicles (AVs), mitigating traffic and congestion while allowing users to spend their drive time relaxing or getting a head start on their day's work. Lunch could be easily ordered from a favorite local eatery and delivered quickly, via drone, directly to any location. And nightlife and daily errands can be transformed by the ability to hail a self-driving car to drive on behalf of the rider or to utilize a robust and responsive public transit network or a form of micromobility to either go to dinner or to the grocery store. Perhaps considering a multimodal approach, individuals might use one or more of these methods daily to maximize efficiency.

The mobility industry, born from a union between the automotive and technology sectors, is revolutionizing the way the world moves people, goods, services, and data in the twenty-first century. Within this movement, we have seen a wave of changes as a result of technological advancements, globalism, and consumer trends, creating new market segments and new options for people everywhere.

Technology has continued to grow, and this is largely attributed to the male workforce, which has a legacy of dominating Western culture. It is no secret that both the technology and automotive industries are prominent spaces for the legacies of "boys' clubs" as both male leadership and staffing continue to dominate their respective sectors and adjacent industries.

Although women are almost half of the United States (U.S.) labor force, they represent fewer than one-quarter of the automotive workforce.[1] That number dwindles to a fraction of employees when looking for women of color in the industry—Black women account for 5.6%, Asian women account for 1.8%, and Latina women make up 3% of the workforce.[2]

According to a 2019 *USA Today* study, of the 11 major automakers, less than a quarter of their U.S. executive staff (vice president or higher) were comprised of women. Apart from General Motors (GM), where 6 of 11 global board members were women at the time, no automaker achieved equal representation on their board of directors. For the majority of the 11 automakers—Fiat Chrysler, Nissan, Honda, Hyundai-Kia, Jaguar Land Rover, and Toyota—an average of 4 in 5 board members were men.[3]

To make matters worse, the industry has difficulty attracting women. In a survey of women in various industries, respondents reported that the automotive industry was one of the least successful at attracting and retaining women.[4] Women cited multiple reasons for avoiding a career in the industry, including an unattractive environment for women (65%), a lack of work-life balance (59%), and a lack of flexible schedules (46%).[5]

This experience is not unusual when examining the world's most powerful and influential companies outside of automotive as well. Although GM has Mary Barra at the helm as the first female Chief Executive Officer (CEO) of an automaker, and there are more women running Fortune 500 companies today than at any point in the 63-year history of the *Fortune 500* rankings,[6] women represent only 6.4% of the Fortune 500's leadership ranks.

What is more demoralizing is that men do not always see the problem. When asked about the extent to which minorities, including women, are represented within

---

[1] U.S. Bureau of Labor Statistics, "Table 3: Employment Status of the Civilian Noninstitutional Population by Age, Sex, and Race, 2019," Current Population Survey, Household Data Annual Averages, 2020; U.S. Bureau of Labor Statistics, "Table 18: Employed Persons by Detailed Industry, Sex, Race, and Hispanic or Latino Ethnicity, 2019," Current Population Survey, Household Data Annual Averages, 2020.

[2] U.S. Bureau of Labor Statistics, "Table 2: Employed and Experienced Unemployed Persons by Detailed Industry, Sex, Race, and Hispanic or Latino Ethnicity, Annual Average 2019," Current Population Survey (unpublished data), 2020.

[3] https://www.usatoday.com/story/money/2019/11/11/auto-industry-gender-diversity-gm-ceo-mary-barra/2513383001/

[4] Deloitte and Automotive News, "Women at the Wheel: Recruitment, Retention, and the Advancement of Women in the Automotive Industry," October 2018, 5.

[5] Deloitte and Automotive News, "Women at the Wheel: Recruitment, Retention, and the Advancement of Women in the Automotive Industry," October 2018, 6.

[6] http://fortune.com/2017/06/07/fortune-500-women-ceos/

their company's leadership teams, 60% of male automotive professionals felt these groups were sufficiently represented or overrepresented.[7]

Despite the patriarchal pedigree and changing expectations and biases over women's professional and personal pathways, the mobility industry is seeing a growth movement in female leadership. This trend exists at all levels. Women are taking their seats in boardrooms, standing alongside machines in manufacturing plants, leading movements in organized labor, testing new technologies on roads, and designing some of the best-in-class models for our future vehicles.

Even more so, as we see numbers rising in relation to gender in the workforce, we are starting to have necessary conversations about gender diversity. Though this introduction contrasts experiences of women against the experiences of men, we recognize that this conversation must include people that identify outside of male and female. As we see women take their seat at the table, we need to make sure there is room for all genders. The intention of this book is to take a step in the direction of a completely embedded gender-diverse culture in these industries as well as others. We have to stop existing and endorsing in a world that limits by two genders.

Women are working every day, often behind the scenes. They are integrated throughout the mobility industry at large, designing and engineering some of the most innovative technologies and platforms that are driving the future of how goods and people move—and the spaces, communities, and roads they move along. Women have broken into a new space with room to merge the foremost automotive and technology elements, design new systems, and create modes of transit that have never been seen before.

Henry Ford, the early 1900s industrialist and father of the modern automobile, famously said, "If I had asked people what they wanted, they would have said faster horses." As legend has it, this was his reference to Ford Motor Company's development of the automobile to help people travel faster and more efficiently. It is a common belief that Ford used what we would now consider design thinking to interpret the people's desires and innovate a solution they did not even know they needed. It is not just about designing a car or a road; it is about designing a future for everyone.

It is interesting, however, to consider how the interpretation of consumers' hopes and dreams of faster travel in the 1900s would have translated had there been a more diverse sounding board rather than just the opinion of one man representing a singular demographic. How would the next generation of transportation have looked if there had been a more holistic view of the population and what their needs truly were at the time? Would it have been different if women were part of that innovation process? Perhaps women were a part of the process but were not given credit or asked for their input—and unfortunately, it is becoming more and more evident that so much of our world has been designed and developed from a limited or single perspective.

Women bring a unique perspective to the table, and it is undeniable that we are all better off when women and girls are able to make their full contribution. Today women represent a huge segment of economic power and offer important consumer insights that are commonly overlooked by their male counterparts. There is plenty

---

[7] https://www2.deloitte.com/us/en/pages/manufacturing/articles/diversity-women-in-the-automotive-industry.html

of evidence that increased participation by women improves the innovation performance of organizations and societies. In fact, research shows that diverse, inclusive teams are more innovative,[8] and diverse companies are more profitable.[9]

By having diverse representation—dare we say, equal representation—at the table, the mobility industry will continue to broaden its innovation and performance and ensure that new products, processes, and methods of transit meet the needs of the whole population, not just the male half. This shift should not be applied exclusively to two genders, but rather to all genders in a space that has been historically dominated by one.

Throughout our early careers—spent working alongside automakers, suppliers, technology providers, infrastructure planners, venture capitalists, technology startups, and transit advocates—we have had the utmost privilege of meeting and working alongside women whom we saw as pioneers and artists creating the future.

While many women spend their careers doing this work, they rarely are recognized publicly for the monumental impact they have on the lives of people in their communities and around the world. This applies not only to public recognition; there is also a lack of internal recognition and personal recognition. Multiple times in the development of these stories, we heard from women that they were "just doing their job" or that it was not "noteworthy." Through the chapters in this book, the gravity and progress made by many of these women display just the contrary.

When examining speaker panels and presentations at industry events across the country, we found that women were once again in the minority when it comes to public speaking engagements. Because of this, Trucks, a venture capital firm, has begun building a crowd-sourced speakers bureau of female leaders in transportation.[10] Thought leaders we see in the media are also overwhelmingly male. However, until event organizers, media, and company executives commit to diverse representation, the female masses will remain in the shadows.

The historic movement of female leadership on the heels of the incumbent male dominance needs to be told. But how? This was a question we asked ourselves, but without resolution. Before making the commitment to author this book, we struggled with how we were going to overcome a systematic gender bias that has been ingrained in this historic industry for generations in order to uncover the women who had been pushed aside, along with their innovative projects. Where do we even start to dismantle the auto and tech patriarchy in which we find ourselves?

This project started on LinkedIn with a simple post: "I am turning to LinkedIn to crowd-source women-led mobility projects, deployments, and/or initiatives as well as automotive and mobility product development across the country." Within 24 hours the post had been viewed more than 13,000 times, spurring an overwhelming response from people across the country and around the globe. Through the responses, we were directly introduced to nearly 100 women and their projects, and within

---

[8] https://www.scientificamerican.com/article/how-diversity-makes-us-smarter/

[9] https://www.mckinsey.com/business-functions/organization/our-insights/delivering-through-diversity?cid=other-eml-nsl-mip-mck-oth-1802&hlkid=5bb59d87545a4618a8ec01d54f62dcc3&hctky=2361563&hdpid=cf69bc13-2f25-43e6-8c6d-793294bbe44e

[10] https://www.trucks.vc/blog/emerging-women-startup-leaders-in-transportation-mobility-autonomous

months, interest in what we were doing with this information grew rapidly. With the demand increasing, we launched a website where anyone could submit themselves or nominate someone else to have their project featured in this book.

The interesting part about this data-gathering experiment was that nearly all the recommendations and nominations were either made by other women or were submitted by the women themselves—further evidence that the female segment of this workforce is not receiving explicit support from their male counterparts and may not be held in high esteem by the men in their organizations. No wonder women struggle to climb the ranks and break that glass ceiling when the work they have done is dismissed by the primarily male leadership.

For us, the message was clear: *these stories were now our responsibility.*

As two women working in the mobility space and as communications professionals and storytellers by trade, we felt it was our duty to elevate and share these stories and to celebrate everything women bring to the table. It became our purpose and mission to create a platform through which to recognize female thought leaders and further showcase diverse perspectives and ways of thinking and its impact on society. It is important to note that we both are two millennial, white, straight women—and though we were committed to providing a diverse perspective of the current state of the industry, we recognize that our perspective still has limits. This book will not be the "cure-all" to this legacy issues, but we believe this conversation and these stories are one step in the right direction.

We spent more than a year researching and tracking down some of the best and brightest minds in the mobility industry across the United States. During this project, we often found ourselves in an interesting position—one in which we had to convince women that they were worthy of recognition and that the projects they had spent their careers building were, in fact, important. In some cases, no one had ever recognized their projects, let alone given them a platform they could use to help others create a meaningful impact in their communities.

It is our goal in bringing together this collection of stories that more awareness be brought to their projects and to the movement of female leadership in the mobility industry. This project is meant to challenge the status quo and elevate the voices of women. Although this book will illustrate the immense effect women have made in every corner of the industry, we want to be very clear: *This book is not just for women.*

This book is about much more than the nearly 150 women featured, and it is definitely not meant to solely be a female-empowerment narrative. From the design and engineering of industry-changing technologies and products, the policy and funding that allows for the further innovation and infrastructure development, and placemaking that makes it all possible, we have done our best to showcase the full and dynamic spectrum of the mobility industry and how it has the power to change nearly every aspect of our lives for the better.

The chapters of this book identify 11 vital pillars of the mobility industry, including awareness and community advocacy, design and engineering, funding, infrastructure, marketing and communications, mobility on demand, placemaking, policy and legislation, sustainability, talent and education, and other technology innovations. Those chapters are followed by a "sign of the times" chapter, which gives a behind-the-curtains look at what happens when crisis occurs and how transportation networks and solutions responded to the COVID-19 public health crisis that has lasted longer than it took us to write this book. Each section digs into the importance of each vertical and discusses the role it plays into the growth and evolution of the mobility industry as a whole. The book includes more than 40 case studies based on real-world projects, deployments, engineering feats, and other innovative transportation solutions—which just so happen to be led by individuals who identify as female.

The pages of this book are meant to serve as a collection of industry best practices and provide examples of how anyone, regardless of gender, can create safer and more sustainable and equitable transportation in their own communities. As you read through the book, you will find case studies that illustrate The importance of regional transit and complete streets programs; How inclusive design matters when serving a community; How tribal communities are implementing transportation programs that best meet the needs of their people; The role of public perception on electric vehicle (EV) adoption and alternative commute modes; The need to focus on workforce development and talent attraction to drive the future of the industry; And much more.

Take this as an opportunity to learn from these world-class projects. Use these stories as inspiration to create monumental change to transportation in your own community. And consider this your call to action to be a champion for women in mobility.

1

Awareness and Community Advocacy

yamasan0708/Shutterstock.com

Women have been at the forefront of transportation community leadership for decades, passionately advocating for changes that saved lives, prevented injuries, and increased mobility for all. Many of the most successful transportation safety campaigns were started by women who were quite simply fed up with the carnage on our roads. The key to success is to connect with road users and demonstrate the project benefits with real-life examples that have meaning for them. For any project or safety campaign to succeed, the engagement needs to be consistent and high-level. That's how you build credibility and win the community's support.

—Nicole Nason, Chief Safety Officer and Head of External Affairs, Cavnue

One of the most important and key indicators of the success of next-generation mobility innovations comes down to public acceptance and adoption. Unless the public is informed and eager to use new mobility options such as bikes, scooters, and other micromobility options, platooning trucks; aerial drones; shared and mass transit; and connected and AVs, the industry will struggle to provide its full value.

These technologies truly need to be both accepted as safe and a value-add by communities to be successful at scale.

Let us take AVs for instance. AVs promise a range of potential benefits, including reduced crashes; greater independence for non-drivers, seniors, and people with disabilities; increased productivity for passengers; and reduced congestion. However, a recent study from Partners for Automated Vehicle Education (PAVE), a coalition of industry players and nonprofits aimed at improving the public's understanding of AVs, shows that Americans are skeptical of the technology. Forty-eight percent of respondents say they would never get in a taxi or ride-share vehicle that was being driven autonomously; however, 60% say they would have greater trust in AVs if they understood better how the technology works—proving that education and awareness could be the key to changing perceptions and attitude, therefore increasing the rate of adoption.

"The results of this survey confirm that autonomous vehicles face major perception challenges, and that education and outreach are the keys to improving trust," said Tara Andringa, Executive Director of PAVE. "These insights provide both motivation and direction to our effort to confront this educational challenge."

The same can be said about public transit and people's willingness to use it. Although public transportation is 93% more cost-effective than maintaining and operating a private vehicle, more sustainable and can support the reduction of the nation's carbon emissions by 63 million metric tons annually, and is 10 times safer per mile than traveling by private vehicle, Americans still tend to opt for other means of transportation, especially in places with limited transit options. In a 2020 study (pre-pandemic) by TransitCenter, a transportation research and advocacy organization, 50% of riders in metropolitan areas with mature public transit systems say they are using transit less frequently than two years ago and are instead turning to ride-hailing apps and car ownership. Today 76% of Americans drive themselves to work, and transportation advocates argue that the creeping shift from transit to private vehicles isn't good for cities. To curb this shift, cities need to make transit fast, accessible, affordable, and convenient, and there needs to be a concerted effort to both listen to and appeal to the current and future commuter by understanding their needs and demonstrating value.

From micromobility and EV infrastructure to AVs and delivery robots, one thing remains true: *the community must come first*.

There are many tools that currently exist to help support these efforts and initiatives, but there is no single "silver bullet." Each community is unique and special in its own right, making outreach a piece of the transportation puzzle that requires just as much planning, detail, and consideration as the others. If there is one thing that rings true about the value of community outreach and engagement, it is this: the people that call a place home are the true subject matter experts of the places we design streetscapes for; they know how a great car design can make or break a busy schedule; they are the authors of the stories we tell through branding; they are the future of this industry.

The projects featured in this section outline the critical importance of listening and understanding the needs of the communities in which they operate and ensuring services are actually solving problems and providing value.

Each demonstrates a community-first approach that deploys transportation and mobility innovations that are intentional, balanced, and trusted by the people they were created to serve.

Census Transportation Planning Products Program

Penelope Weinberger, Transportation Data Program Manager, American Association of State Highway and Transportation Officials (AASHTO) Washington, District of Columbia (DC)

Background

Operating previously ad hoc with the decennial census since 1980, the Census Transportation Planning Products Program (CTPP) was officially established as a permanent, ongoing program in 2012 by the American Association of State Highway and Transportation Officials (AASHTO). The CTPP is a cooperative program funded by state Departments of Transportation (DOTs), providing open data access to the transportation planning community. The CTPP additionally funds and conducts transportation research and provides training and technical assistance. The transportation planning community has particular needs for data centered on where people live, work, and travel. These professionals rely on census-related projections for multi-year and long-range planning efforts. The CTPP uses the American Community Survey (ACS), an ongoing survey that releases data annually, to develop transportation-specific tabulations that include the following:

- Residence-based data.
- Workplace-based data.
- Commuter flows from home to work, also known as Journey to Work (JTW).

The ACS data is released in one- and five-year aggregations according to a population threshold. The CTPP takes that data and reframes it to provide geographic patterns of the JTW commute as well as personal, commuting, and household characteristics.

This data is used to validate travel demand models, develop vehicle ownership models, analyze demographics of communities, and perform equity analyses and can be applied in many more applications, while also examining household life-cycles and providing population forecasts. These datasets are critical to transportation projects as feasibility, design, funding, implementation, and other planning elements are considered prior to the study, design, or construction phases of such projects.

The transportation portion of census data, which includes data at small geographies and flows from home to work, crossed with worker demography. The CTPP data is a baseline dataset in many transportation models. It is used by most MPOs and states, along with many users in nontransportation industries for planning, modeling, analysis, and, in 2020, COVID-19 movement tracking. Additionally, it provides a snapshot of the nation's workforce mobility in a comprehensive way. The CTPP, however, is more than a dataset; it is an entire program of training, research, and policy dedicated to understanding, using, and incorporating census transportation data.

Goals and Objectives

The initial goal of the program was to develop a unified buyers' pool to purchase specialized data tabulations. Today the CTPP purchases the data and produces a free and open data portal and Geographic Information System Mapping (GIS) platform. The CTPP additionally sponsors a conference on the matter every few years. The technical leadership supports a community of practice and advocacy for the continued quality collection of transportation data provided by the U.S. Census.

The CTPP is a resource that transportation planners can rely on when working on new projects and improvements to existing infrastructure. The CTPP exists to provide enhanced and targeted datasets from the ACS. These data are the foundation of corridor and project studies, environmental analyses, and emergency operations management.

Approach and Challenges

The CTPP facilitates the provision of the gold standard of planning data: a special tabulation. The software provides a variety of dataset formats; data can be downloaded in CSV, Microsoft Excel, shapefile, and tab formats. The website also allows analysis previews within the software. Datasets can be accessed as early as 1990.

The CTPP uses custom terminology to describe certain ways that the datasets are analyzed. A Transportation Analysis District (TAD) is a unique designation to the CTPP that refers to a population threshold of 20,000. This was added so that the ACS three-year tabulation can provide data limited to geographic areas above the threshold.

Results

As of 2021, there are more than 1,500 registered users of the most recent dataset. The CTPP has more users because it can be fully accessed without signing in, but registration is required for saving customizations.

The program continues to adjust to changes on the census as it evolves. One example of platform modifications is that with the end of the three-year ACS datasets, there are no more three-year CTPP tabulations. Additionally, TAD-level and traffic analysis zones (TAZ)-level geographies are prohibitively expensive and will be permanently discontinued as of the 2017-2021 dataset.

Some of the reports and research papers developed from these data include:

- Commuting in America; Developing a National Report from National Survey Sources, Weinberger, 2016[1].
- The Commuting in America Series: three volumes from 1989, 1996, 2006; 17 briefs in 2013; an ongoing series beginning release in 2021 (two published, four anticipated for 2021).
- National Datasets; How to Choose Them, How to Use Them, Weinberger, 2016[2].
- Origin-Destination Trips by Purpose and Time of Day Inferred from Mobile Phone Data, Alexander et al., 2015[3].
- A Multi-Scale Analysis of Urban Form and Commuting Change in a Small Metropolitan Area (1990-2000), Horner, 2007[4].
- Understanding the Role and Relevance of the Census in a Changing Transportation Data Landscape, Erhardt and Dennett, 2017[5].
- Vehicle Availability and Mode to Work by Race and Hispanic Origin, 2011[6].

About the Team

Penelope Weinberger is AASHTO's Transportation Data Programs Manager. Weinberger is the liaison to the Committee on Data Management and Analytics (CDMA) and runs the CTPP. Weinberger has worked for AASHTO since 2009. In her tenure, she brought the CTPP from an ad hoc to an ongoing program and was the first AASHTO staff assigned to the newly formed CDMA. In the development of the AASHTO reorganization that created CDMA, Weinberger authored the seven core data principles that guide the work of data users. Weinberger was previously with the Texas A&M Transportation Institute working on the Travel Model Improvement Program and Cambridge Systematics where, among other things, she managed analytics for the "Bottomline Reports to Congress." Before moving to the transportation industry, Weinberger had a colorful career and can mix a Manhattan and froth a cappuccino with equal skill. Educated at the University of Illinois, Chicago (and by life), she is an avid contra dancer, cyclist, and bridge player.

---

[1] https://ctpp.transportation.org/wp-content/uploads/sites/57/2018/11/CIA_Penelope_2016.pdf
[2] https://ctpp.transportation.org/wp-content/uploads/sites/57/2018/11/NationalDataSet_Penelope_2016.pdf
[3] http://humnetlab.mit.edu/wordpress/wp-content/uploads/2010/10/ODs_TRC15.pdf
[4] https://link.springer.com/content/pdf/10.1007/s00168-006-0098-y.pdf
[5] https://ctpp.transportation.org/wp-content/uploads/sites/57/2018/11/ConferencePaper_Keeping-Census-Relevant_pdf.pdf
[6] https://www.census.gov/content/dam/Census/library/publications/2015/acs/acs-32.pdf

Closing EV Workforce and Ownership Gender Gaps

Erika Meyers, EV Love, Founder Global Senior Manager, Electric Vehicles, World Resources Institute
Washington, DC

Background

When considering the gender gap, people often are quick to think about income-based disparity. This disparity also exists among genders when it comes to ownership of hybrid and EVs. Surveys have shown that men are significantly more likely to purchase EVs than women. Women are responsible for 80% of purchasing decisions for households and purchase more than 50% of new vehicles but only represent 30% of EV purchases. After researching this trend, Erika Myers, the founder of EV Love, discovered that marketing messages are often targeted to men and are predominantly focused on performance and technology advancements. Studies show that women prioritize safety, value, and dependability, and because EV technology is new (and therefore perceived as a riskier investment) and more expensive than conventional vehicles, women are more likely to pass on the purchase.

An additional layer of disparity is that very few women have decision-making authority in the automotive industry. Only 7% of executive positions are held by women—in the top 25% of automotive brands. Further, fewer than 25% of all automotive jobs are held by women. The lack of female gender diversity throughout the automotive industry has residual effects on how EVs are marketed, designed, and sold.

Goals and Objectives

EV Love (www.electricvehiclelove.com) was created to help bridge this gap by providing a safe space for women to learn more about EV technology and understand the benefits of going electric. The website combines factual research with pithy and entertaining commentary intended to do three things:

1. Help women cut through the clutter of online EV information.
2. Inform the automotive industry about the unique needs of women.
3. Encourage more women to consider employment in the EV industry.

Approach and Challenges

By leveraging an extensive background in clean energy and alternative transportation fuels (including EVs), Myers is creating another space to amplify the voices of those who are EV consumers and/or are employed by the EV industry.

After a decade of observing this gender gap in EVs and clean energy technology, Myers took advantage of the pandemic to bring the EV Love website to life. The official research started in March 2020, with the public launch in November of that year.

Myers stated, "Most people would probably do this type of project once they had secured funding, but I did it as a passion project with my own money to raise awareness of this critical issue." While the mission of EV Love is the forefront of the project, it was also intended to present the stories and pictures of real people who are missing entirely from the EV community.

Results

EV Love aims to influence the market at large through content that increases awareness and inspires change, leading to more households behind the wheel of an EV and more gender diversity in the automotive and clean energy workforce. EV Love will continue into perpetuity until the gender gaps in EV sales and workforce are a nonissue.

About the Team

Erika Myers is an EV subject matter expert at World Resources Institute (WRI) Ross Center for Sustainable Cities. She leads the electric mobility team's research efforts and works with cities across the globe to identify opportunities to electrify transportation, including infrastructure deployment and vehicle-grid integration methods for public transit and other municipal fleets.

Myers has worked for nearly two decades on clean energy and alternative transportation fuels and distributed energy resource topics in government, for-profit, and nonprofit roles. Her background gives her a unique perspective on the opportunity to leverage renewable energy and EV charging to reduce emissions through the widespread deployment of vehicle-grid integration. She walks the talk by owning two battery EVs powered by 100% renewable energy, which she manages through networked residential chargers.

Myers is a frequent speaker at energy industry events and has published dozens of reports and white papers on transportation electrification and clean energy issues. She currently serves on an EV steering committee for the U.S. Department of Energy (DOE) and holds a leadership position in the Women of Electric Vehicles, Washington, District of Columbia (DC) Chapter. Erika previously served as an EV advisor for the Fuels Institute, the National Energy Foundation, and U.S. DRIVE (Driving Research and Innovation for Vehicle efficiency and Energy sustainability). Myers was awarded the 2019 Public Utility Fortnightly (PUF) "Fortnightly Under 40" award for her work on vehicle-grid integration and a PUF Innovator Award in 2018. Prior to joining the WRI, Myers was the Principal of Transportation Electrification for the Smart Electric Power Alliance (SEPA), an alternative fuels and clean energy consultant with ICF International, a renewable energy manager for the South Carolina Energy Office, and a Clean Cities Coordinator for the Palmetto State Clean Fuels Coalition.

Myers holds a Bachelor in Biology from Clemson University and a Master in Earth and Environmental Resource Management with a concentration in renewable energy and climate change from the University of South Carolina.

In her spare time, Myers volunteers for her city's Energy Transition Subcommittee to increase the number of public EV charging and procure electric buses for the local school district. She also mentors young professionals interested in a clean energy career.

> ***Extract:*** "The work I'm doing will hopefully lead to positive changes for my young daughter. As a mother, I am very protective of her and don't want her to experience the gender inequities I have throughout my career. Further, my lifelong devotion to transportation decarbonization will help keep the planet as normal as possible for her and future generations. While I realize that this passion project is a very small step to solve that a much bigger issue, it still makes me feel like I'm making a difference. I hope this work will inspire other women to figure out how they can use their talents to be a positive force for change. Because individually we can't do much, but collectively we can change the world."

Michigan Long-Range Transportation Plan Public Engagement Partnership for Inclusive Participation

Monica Monsma, Public Involvement and Hearings Officer, Environmental Services, Michigan Department of Transportation
Michigan

Background

In compliance with the Federal Highway Administration (FHWA)'s requirements for long-range planning, the State of Michigan Department of Transportation (MDOT) launched their effort in 2018 with the Michigan Long-Range Transportation Plan: Michigan Mobility 2045 (MM2045). This plan is a multiyear effort that intends to identify strategies, vision, and priorities for transportation in Michigan.

It is the policy of the FHWA that public involvement and a systematic interdisciplinary approach are essential parts of the development process for proposed actions [23 (Code of Federal Regulations) CFR § 771.105(c)]. Each state must have procedures approved by the FHWA to carry out a public involvement/public hearing program pursuant to 23 U.S.C. 128 and 40 CFR parts 1500 through 1508. State public involvement and public hearing procedures must provide for:

- Coordination of public involvement activities and public hearings with the entire National Environmental Policy Act (NEPA) process.
- Early and continuing opportunities during project development for the public to be involved in the identification of social, economic, and environmental impacts, as well as impacts associated with the relocation of individuals, groups, or institutions.
- One or more public hearings or the opportunity for hearing(s) to be held by the state highway agency at a convenient time and place for any federal-aid project that requires significant amounts of right-of-way; substantially changes the layout or functions of connecting roadways or the facility being improved; has a substantial adverse impact on abutting property; otherwise has a significant social, economic, environmental, or other effects; or for which the FHWA determines that a public hearing is in the public interest.
- Reasonable notice to the public of either a public hearing or the opportunity for a public hearing, and such notice will indicate the availability of explanatory information. The notice shall also provide the information required to comply with public involvement requirements of other laws, executive orders, and regulations.

The Safe, Accountable, Flexible, Efficient Transportation Equity Act (SAFETEA-LU) Environmental Review Process Final Guidance on coordination and schedule provides information on how lead agencies should establish a plan for coordinating public and agency participation and comment during the environmental review process. It also provides guidance on updating public involvement procedures to help state DOTs determine if they need to update their public involvement procedures pursuant to 23 CFR § 771.111(h).

In addition to these requirements, the state DOTs also must:

- Pursue communication and collaboration with federal, state, and local partners in the transportation and environmental communities, including other modal administrations within the U.S. DOT.
- Seek new partnerships with tribal governments, businesses, transportation and environmental interest groups, resource and regulatory agencies, affected neighborhoods, and the public.
- Ensure that those historically underserved by the transportation system, including minority and low-income populations, are included in outreach.
- Actively involve partners and all affected parties in an open, cooperative, and collaborative process, beginning at the earliest planning stages and continuing through project development, construction, and operations.
- Ensure the development of comprehensive and cooperative public involvement programs during statewide and metropolitan planning and project development activities.

Understanding the purpose and need, proposed alternatives for implementation and construction, and potential impacts of the proposed project as well as the mitigation of environmental impacts are required from the FHWA, and compliance with these requirements restricts federal funding to these projects and their respective agencies.

The success of the public engagement strategy and execution may determine future funding outcomes for the implementation of the transportation projects, and if there is not a good faith effort for engagement, funding can be reduced.

Through Phase I and Phase II of MM2045 public engagement, the project team identified gaps in transportation options, infrastructure, and accessibility. The key takeaways identified in the public engagement period of the plan development will be used to drive the state's next set of transportation and mobility plans. The following mission statement was developed for the plan development:

> In 2045, Michigan's mobility network is safe, efficient, future-driven, and adaptable. This interconnected multimodal system is people focused, equitable, reliable, and convenient for all users and enriches Michigan's economic and societal vitality. Through collaboration and innovation, Michigan will deliver a well-maintained and sustainably-funded network where strategic investments are made in mobility options that improve quality of life, support public health, and promote resiliency.

These requirements are traditionally met with in-person public meetings and conference presentations and by hosting tables at places like farmers' markets, which members of the community are already visiting. Phase II of MM2045 kicked off at

FIGURE 1.1 An American Sign Language interpreter joins on-screen during a 2021 MM2045 meeting to offer an accessible meeting format for participants. She is interpreting an exciting conversation on passenger rail needs being led by Joe Gurskis.

Courtesy of Kristen Shaw.

approximately the same time the height of the COVID-19 pandemic was hitting Michigan and the rest of the country (Figure 1.1). This presented its own challenges, but for MM2045, it required a quick pivot to virtual public engagement in order to stay on track for the timeline.

The success of this engagement was dependent on a unique partnership formed by MDOT and the Michigan Department of Civil Rights, which together helped identify ways in which all Michiganders would be able to participate in the public engagement process, especially now that it was being moved online and in-person engagement was not feasible.

Goals and Objectives

The goal of MM2045 is to build a transparent plan on future infrastructure so that it belongs to all Michiganders. Traditional long-range plans focus on select transportation modes, but MM2045 wanted to take it a step further and include modes such as freight, rail, and marine transportation. Everything is dependent on a strong and modernized infrastructure system, from the economy to emergency response times to the options available for people to get from one place to another. MDOT intended for this plan to serve as a way to capture feedback at every level needed in order for the Great Lakes State to have a vibrant transportation system to meet its unique needs.

These were the overarching goals for the plan development process. The partnership with the Department of Civil Rights presented a secondary set of objectives, led by the main goal of reaching as many Michiganders as possible. This goal, which came from the partnership, refined the way that the initial outreach was planned.

Approach and Challenges

The partnership between MDOT and the Michigan Department of Civil Rights began in summer 2020, with conversations following the publication of the Phase II survey results. The pivot to virtual engagement was not comprehensive enough to provide accessibility to all people. This partnership led to expanded options for surveys, including better compatibility with e-readers, an American Sign Language survey with videos, improved translation capabilities, and options for participants to engage by writing and returning their responses by mail or via phone call. Adjustments were also made so that participants could submit the survey on behalf of another participant if that individual was preferred not to participate independently.

The use of expanded technology allowed the MDOT to extend outreach to a larger and more diverse group of Michigan residents in developing MM2045.

Types of outreach included in the phases of public engagement included:

- An attitudes and perceptions survey to ensure the needs and priorities of a representative sample of Michigan residents are considered.
- Scenario planning workshops to present transportation situations to the public in a realistic context to help determine a long-term vision for transportation in Michigan.
- Collaborative public engagement, including targeted outreach to environmental justice populations and marginalized communities.
- Focused outreach for interdepartmental and external agency collaboration including tribal governments.

The goals of this partnership were accomplished through removing barriers from individuals with disabilities or who identify as deaf, deaf/blind, hard-of-hearing, or have limited-English proficiency, as well as environmental justice populations.

The Environmental Protection Agency (EPA) defines Environmental Justice as "the fair treatment and meaningful involvement of all people regardless of race, color, national origin, or income, with respect to the development, implementation, and enforcement of environmental laws, regulations, and policies." This goal will be achieved when everyone enjoys the same degree of protection from environmental and health hazards, and equal access to the decision-making process to have a healthy environment in which to live, learn, and work. Considerations to include houseless, migrant workers and dispersed populations were also a part of this outreach.

In addition to making sure these populations were included, this partnership led to the identification of strategies that the MDOT can incorporate into the long-range transportation plan that will help Michigan's transportation system remain competitive and economically viable while serving all populations—supporting residents and visitors collectively and individually to increase the quality of life in Michigan. The question to answer was "What strategies can we include in our plan that make life better in Michigan?"

For all public involvement, the MDOT provided multiple pathways to look at the needs of these communities and tailored how they would communicate with them and develop this plan. Some of the tools included were Survey Monkey, MetroQuest, and PDF surveys; mailers; virtual meetings and workshops; in-person meetings; attendance and presentations at other public meetings; e-newsletters; stakeholder involvement with identified community groups, nonprofits, and neighborhood organizations; a project website to serve as a home for all of these communications; a social media and video campaign with videos that had both closed-captioning and American Sign Language.

Results

The partnership between the MDOT and the Michigan Department of Civil Rights was crucial to this level of expansive engagement, and the MDOT expects this partnership to continue on many future projects. Collaboration between different departments is untraditional, but it truly is an example of adaptation when faced with unique challenges posed by the pandemic and also the nature of a long-range plan due to size and demographic variants throughout the state. Social media turned out to be one of the best tools to help meet the needs of this level of outreach, specifically those in environmental justice populations. This partnership allowed the MDOT to engage with more people than previous outreach, resulting in robust datasets to help develop the LTRP strategies and insights with additional contexts.

At the time of this publication, Phase II was still in progress, but the results display the success of the outreach during Phase I and in Phase II.

About the Team

Monica Monsma is the Public Involvement and Hearings Officer for the MDOT. She facilitates public engagement for regional and statewide transportation projects with her expertise in NEPA and FHWA public involvement procedures. Prior to her role with the MDOT, Monsma was Executive Director of the Chelsea Chamber of Commerce. Monsma holds a Bachelor of Arts (B.A) degree in English and history from Western Michigan University.

2

Design and Engineering

Eugenio Marongiu/Shutterstock.com.

> I value my existence by the value I can bring to others. Fundamentally, technology should serve to improve people's lives, and connected and automated technology has great potential to do just that. As a woman in the transportation technology domain, I believe my unique perspective, combined with the unique viewpoints of my colleagues, is crucial to realizing this great potential for the betterment of society.
>
> —Sue Bai, Division Director, Honda Research Institute USA, Inc.

It is important to note that this chapter was originally intended to be titled "Design, Engineering, and Manufacturing," but, as you can see, "Manufacturing" has been stripped away. Women's roles in manufacturing date back to 1914, at the start of World War I.[1] After the men had gone to war, women had to step in and take over the jobs they left behind, specifically on the automotive assembly lines that were converted to manufacture and keep up with the high demand for military vehicles

---

[1] https://www.striking-women.org/module/women-and-work/world-war-i-1914-1918

by the United States and other ally countries. This was the start of women's vital roles in the automotive industry. Later, the U.S. Government deployed the "Rosie the Riveter"[2] campaign, which was aimed at recruiting even more female workers for the defense industries during World War II. Rosie perhaps became the most iconic image of working women—and as they say, the rest was history.

Maybe it is because we find ourselves living and working in the Metro Detroit area, a place known for its manufacturing prowess and its legacy around building cars people love and rely on, but we assumed it would be easy to find a female-led case study that embodied innovations to the assembly and fabrication process. We were wrong.

Not only are women in automotive manufacturing already a small pool, but female leadership within this pool is even smaller. After many pleas on social media and introductions to women working in manufacturing, we spoke with many amazing women; however, we were not able to identify specific projects on which to create a case study. We know there is a story out there that needs to be told. It is one of our biggest areas of discontentment with writing this book that we were not able to uncover a manufacturing story to share. Instead, we focused on design and engineering—two other disciplines that are primarily male dominated.

Despite making up nearly half of the U.S. workforce, women are still vastly underrepresented in the science, technology, engineering, and math (STEM) workforce, which includes engineers and designers. According to the 2019 U.S. Census,[3] women made gains in these disciplines in recent years—from 8% of STEM workers in 1970 to 27% in 2019—but men still dominated these fields as they account for 73% of all STEM workers. Despite the progress made, in the last five years, the National Automobile Dealers Association (NADA)[4] found that only 1 in 5 persons entering technical automotive fields are women.

This lack of representation of women in the design and engineering of automobiles has left us with dire consequences that are only now being considered. The truth is we have primarily been operating within a transportation and mobility landscape that was designed by men for women. For example, until recently crash-test dummies were based exclusively on the "average" male, which put women at risk and subsequently has caused women to be 47% more likely to be seriously injured and 17% more likely to die in traffic crashes.[5] Perhaps had more women been involved in the design and engineering of crash-test safety protocol, more lives could have been saved.

Furthermore, there is a significant lack of gender diversity in urban design, which has impacted the way an entire population of people move about in their communities and accomplish tasks throughout the day. When women designers fulfill a lead role in urban design,[6] we tend to see more well-lit spaces to make walking at night safer; wider sidewalks so pedestrians can navigate with children, wagons, and strollers; and the installation of more ramps at stairs and crosswalks to accommodate people

---

[2] https://www.history.com/topics/world-war-ii/rosie-the-riveter
[3] https://www.census.gov/library/stories/2021/01/women-making-gains-in-stem-occupations-but-still-underrepresented.html
[4] http://www.dealeron.com/wp-content/uploads/Jody-DeVere-NADA-Workforce-Report-2016.pdf
[5] https://www.theguardian.com/lifeandstyle/2019/feb/23/truth-world-built-for-men-car-crashes
[6] https://www.bloomberg.com/news/articles/2013-09-16/how-to-design-a-city-for-women

with strollers or individuals using a walker or wheelchair. These elements are often overlooked by male designers.

Recently, we are seeing a rise in women-centered design, which is the acknowledgment that women pose an incredible market opportunity and implores all designers, innovators, and investors to drive the market in ways that are cognizant of women's needs. And it makes sense considering that women's spending drives the world economy at roughly $20 trillion annually.[7]

Ultimately, it comes back to having diverse perspectives at the table. In this chapter, the case studies showcase a wide breadth of female-led innovations that take into account all types of end users regardless of gender, age, abilities, and socioeconomic status to ensure the best and safest transportation experience—from interiors and access to rethinking how people respond to their transit experience.

Creating Accessible AV Transportation for Those with Disabilities or Impairments and for Seniors

Erin McCurry, Product Manager, Accessibility, May Mobility
Tara Lanigan, Head of Policy and Advocacy, May Mobility

Background

In the United States, nearly one in five people have a disability.[8] One of the early promises of AVs was to bring transportation independence to those who cannot drive themselves, including some people who use a wheelchair, experience blindness or deafness, people with chronic health conditions, and seniors.

Driverless transportation is an opportunity to revolutionize personal transportation that offers additional mobility options to people with disabilities. However, the timeline for the deployment of Level 5 self-driving vehicles (full driving automation that does not require human attention) is highly debated among analysts and industry experts, and beyond that, the timeline for mass availability is even more unclear.

However, there are some AV companies operating on the roads today. Many are focused on pilot deployments and testing technology. May Mobility, a self-driving shuttle company with service in several large metropolitan areas, set part of their mission to fulfill the promise of bringing transportation independence to more people.

---

[7] https://medium.com/menstrual-health-hub/women-centered-design-the-future-of-innovation-investment-bed1a021e542
[8] https://www.census.gov/newsroom/releases/archives/miscellaneous/cb12-134.html

Goals and Objectives

From day one, May Mobility was dedicated to providing shared mobility solutions to the general public. This means they needed to ensure that everyone was able to use their service and that their shuttles were accessible for people with disabilities.

To do this, May Mobility set out to better understand the needs of riders with disabilities in order to improve the accessibility of their products and services. May Mobility prioritize connecting with people and groups throughout their pilot programs to ensure they have input from end users from development to deployment.

Additionally, May Mobility wanted to use this platform to draw attention to the importance of inclusive design in transportation, the involvement of disability advocacy groups in the process, and increasing safe, sustainable, and accessible transportation options for people with disabilities.

Approach and Challenges

May Mobility held multiple feedback sessions with riders with disabilities at each of their launch sites in 2019, uncovering new adjustments that could be made to their shuttles with each session. Most vital was the disability awareness workshop led by Feonix Mobility Rising, held in Grand Rapids, Michigan, that fall. This event brought together disability advocates and people with disabilities to understand the challenges that all people face, especially when using public transportation. Additionally, the workshop made space to have conversations surrounding AV technology and the hopes and fears of people with physical disabilities around integrating AVs into their commute habits. Learnings from the workshop were also used to create a training plan aimed at educating Fleet Attendants—the onboard safety drivers—and site teams on how to safely assist and secure passengers.

This event was also an opportunity for May Mobility to workshop and understand challenges and needs of people with nonphysical disabilities, such as cognitive and sensory disabilities, as the company began working on ways to make the service more accessible to people with other needs.

That same fall, the company began working with the National Federation for the Blind to give a demonstration at their annual Michigan chapter meeting and gather insights and feedback. As a result, May Mobility started developing sounds to notify passengers of vehicle events such as when it is approaching a stop or departing and when it is safe to exit the vehicle.

The team also spent considerable time researching and reviewing legal requirements and the Americans with Disabilities Act (ADA) and technical limitations and analyzing rider feedback as it developed an actual product.

From the beginning of the project, the team understood that accessible design meant much more than the inclusion of a ramp in their vehicles. However, by conducting rider interviews and collecting feedback on their product, the team uncovered even more barriers to remove in order to make AVs usable by all. One big component on the journey is properly securing a wheelchair or mobility device in an AV. Often, bus drivers assist wheelchair users with securely restraining their wheelchairs for the duration of the trip. However, there was not a great off-the-shelf solution for May Mobility to implement to easily replace that interaction. The team soon began working with the University of Michigan Transportation Research Institute, which was working on an automatic wheelchair securement system, a device that can automatically anchor a wheelchair to a vehicle to prevent the wheelchair from rolling around during travel.

May Mobility's approach to this challenge aimed to make the development process evolve around the people it aimed to serve. By involving people with disabilities, holding workshops, and gathering feedback at demonstrations, the team is building a foundation in which to continue to improve as it goes to market.

Results

Currently, May Mobility have wheelchair-accessible shuttles available at all of their operational sites that are open to the public. They have partnered with local disability advocacy groups to develop a playbook on accessible AV public transit.

May Mobility have been an active participant in a dialogue around making transportation purpose-built and accessible. With everything accomplished, the team recognizes this project is not over, and there is still more progress to be made.

Next Steps

As an AV company, May Mobility's goal is to be able to provide a fully driverless transportation service to everyone. Knowing there is still more to learn and do to truly accommodate and service people with disabilities, the company applied for the U.S. Department of Transportation (USDOT)'s Inclusive Design Challenge to further fund research and deployment of their accessible self-driving shuttles. In January 2021, May Mobility was announced as a semifinalist, and they are continuing their work to make sure autonomous public transit is built for everyone.

Ultimately, May Mobility hopes to continue to encourage other AV and transit companies to invest in accessible design to bring safe and independent transportation to more people.

About the Team

May Mobility is a self-driving shuttle startup based in Ann Arbor, Michigan, that is focused on supporting cities in becoming greener, safer, and more accessible. Two team members focused on bringing May Mobility's mission to life are Erin McCurry, Accessibility Product Manager, and Tara Lanigan, Head of Policy and Advocacy.

Erin McCurry is Product Manager of Accessibility at May Mobility, where she defines and executes on the company's accessibility roadmap. She previously served as Technical Program Manager, leading releases of new software and hardware features to the May Mobility fleet. McCurry first joined May as Test Engineer and helped develop standards and procedures for testing the autonomy stack. Prior to joining May Mobility, McCurry worked at the U.S. EPA, where she analyzed emissions and driving trend data. She graduated from Eastern Michigan University with a Bachelor of Science (B.S.) in Electronics Engineering Technology.

Tara Lanigan is Head of Policy and Advocacy at May Mobility, where she previously led Business Development and Customer Success. She steers May Mobility's policy efforts on workforce development, accessibility, and safety. She has worked at Michigan-based startups for the majority of her career and received her bachelor's degree in political science from the University of Michigan. She serves on the board of Women in Mobility-Detroit.

Sustainable Material behind Fashionable, Handcrafted Interiors

Adelaide Begalli, Lead Color, Materials, Finish Designer, Karma Automotive
Orange County, California

Background

What is it about California that attracts innovators and adventurers, the self-possessed and daring? They follow a beckoning West to find free-thinking individuals like themselves, who over generations built this place into a capital of creative and engineering excellence. They find natural wonders unlike anywhere else in the United States, from the redwood forests of Northern California to the lush, craggy coast of Big Sur, down south through Hollywood, and into the Baja California Peninsula. The Pacific Coast Highway (PCH) runs the length of the state from San Diego to San Francisco, dancing along the shoreline of Earth's largest body of water. Driving along California's idyllic coast, life suddenly feels less encumbered—an enchanting escape for those who find freedom behind a steering wheel. Experiencing the coast to its fullest requires the right car, and that car was envisioned by Karma as the SC1 Vision Concept roadster.

Karma Automotive is a producer of luxury electric vehicles founded in 2014 and headquartered in Irvine, California. The automaker is known for its environmentally friendly and immaculately designed sedans. Not only was the exterior of the SC1 a piece of art, the cockpit's interior—which is often an afterthought by consumers—was just as thought-provoking, featuring unconventional engineering and trimmings.

Goals and Objectives

After outlining the mission and proportions of their concept car, Karma announced an internal design competition for designers and craftsmen to put together proposals for how they thought Karma's first-ever concept car should look and feel. For the interior finishes, one could imagine themselves behind the wheel, driving north on the PCH toward the San Francisco Bay. What would you see? How would you feel? The interior would be a direct reflection of the sensations experienced in that moment.

Most important, SC1 design would need to be a pioneer in the way materials were used as well as in the way Karma expressed itself. Overall it would need to present a new statement for the future.

Approach and Challenges

As the designer studied new fascinations and imagined what it would be like to be behind the wheel of the Karma SC1 Vision Concept driving north on PCH toward the San Francisco Bay, they considered what feelings, touches, and sensations would be experienced in that environment. The designer started collecting pebbles, leaves, pieces of wetsuits, cashmere, and whatever else came to mind while envisioning the experience.

The interior would be inspired by the earth and rain using never-before-seen materials and finishes that gave the Karma SC1 Vision Concept vehicle personality and distinction. The designer wanted to ensure Karma was a leader in sustainable material design, utilizing locally handcrafted fashionable materials when possible.

Results

The SC1 roadster's use of unusual materials, decorative finishes, and color is spectacular. The vehicle's exterior paint is a one-off, fluorescent orange with flakes of violet mica that contrasts nicely with the cabin, which subtly fades from blacks to dark blues. A thin, color-changing accent light runs the width of the cabin and continues down along the door panels. Rubber neoprene is used on some surfaces—not too dull, not too shiny, inspired by California's surf culture—bonded with gentle Italian cashmere. The varnished, waxed edges of SC1 Vision's seats are threaded with impossibly thin, light-emitting diode (LED)-lit fiber, and the sides of the soft leather seats are finished in elegantly laced hides. Several interior structures, including the center, are made from chopped carbon fiber flakes; each piece is molded by hand.

The designer passionately wove California character and style into the design while sourcing local and sustainable materials that are both rich and vibrant with color while also delivering a low-impact solution that is also sustainable.

The Karma SC1 Vision Concept debuted in 2019 to much fanfare. The success and praise over the interior of the SC1 has led to the beautiful execution of new product-line interiors of the Revero GT SCI concept car and future models with a consistent handcrafted quality and very rare levels of customization, which adds a bold feminine touch to the Karma brand.

About the Team

A trio of our designers were chosen to develop their designs, work together, and lead the Karma SC1 Vision Concept project: Concept Design Manager Jacques Flynn; Adelaide Begalli, Lead Designer for Colors, Materials, and Finishes; and Andre Franco Luis, Director of Interior Design.

Adelaide Begalli has shown a dedication to material design and fashion since before starting her career in Milan, Italy. She grew that appreciation during her first role with Toyota Motor Company in Japan, working with traditional weave and stitching processes that outfitted the interiors across the model lineup. When she moved to the United States in 2014, Begalli applied her efforts to automotive leather with Garden State Tanning, where she learned extensively about the trim and dye process used in the interiors of high-end brands such as Cadillac and Aston Martin. Begalli knew there were more sustainable and handcrafted ways to produce luxury products for automotive. She found her place with Karma Automotive in 2016, and as of 2021, Begalli is Senior Designer for Color and Materials at Tesla. Begalli studied Fashion and Textile Design at the Istituto Europeo di Design in Milan.

A Smartphone-Human Hybrid Intelligence System to Measure and Shape Happiness, 2020

Yingling Fan, Professor, University of Minnesota and CEO, Daynamica Inc.
Minneapolis, Minnesota

Background

Transportation is an emotional landscape that people experience regardless of their mode choice. Emotional well-being during a trip is an essential aspect of promoting human-centered transportation planning practices. Yet traditional transportation planning practices have often placed greater emphasis on travel speed and efficiency than on people's emotional experiences during a trip, even though many service providers participate in some level of customer service outreach. That outreach tends to focus on schedule, service distance, and cleanliness of both stops and vehicles themselves. The lack of emphasis on people's transportation experiences is partly due to measurement challenges: people's emotional experiences are fleeting and difficult to recall and quantify.

Modern humans are already tracking every step they take, calories consumed or burned, sleep data, headphone volumes, heart rate, and hundreds of other metrics that have become available to us with the advancements in wearable technology. Many elements of our daily lives are tracked and measured, allowing humans to optimize ourselves by doing things like walking more, eating better, or spending less time in front of screens. While many of these technologies connect us to numbers that may indicate or result in our happiness, they are rarely used as a solution for just that.

Examining user experience in transportation services offers a new lens to understanding how our transportation affects our mood and quality of life. Traditional travel data collection methods can detail activity, destinations, and travel time but can be time consuming, expensive, and too much of an effort for real-time results to be used to improve transportation experiences in an immediate or short-term time frame. Real-time data and adaptations are becoming the expected standard for practicality for our increasingly time-stressed populations.

Understanding these challenges and the opportunities this application and data could provide a solution by presenting a unique position where technology can meet emotion for increased optimization (and happiness) with the mobile application Daynamica. Daynamica is an innovative, research-grade smartphone app that can collect, process, and report data on people's everyday activities and trips with limited interaction from the user. The application then allows the user to view and annotate that data. Not only does this application allow for more accurate data collection than other solutions for measuring human experience during activities and trips but it does so more quickly and at a lower cost.

Goals and Objectives

The goal behind the development and implementation of the Daynamica app is to invite transportation service providers, DOTs, municipalities, and other entities to understand the relationship between emotion and mobility. Organizations and providers can deploy this technology to their customers and users in order to better understand how different modes can affect user experience and overall happiness in a place or during certain times. Daynamica exists as a brain trust focused on inventing new tools to systematically measure people's travel behavior and experience throughout the day with the objective of using those measurements to find ways to promote emotional well-being during trips.

Approach and Challenges

Over the past 10 years, Dr. Yingling Fan and her collaborators invented a smartphone-human hybrid intelligence system, namely, Daynamica, in which smartphone sensing collaborates with human input to generate more accurate and comprehensive data related to daily travel behavior and well-being. Specifically, Daynamica pairs advanced data-mining and machine-learning techniques with a sophisticated user-interface design to allow two-way interactions between smartphone- and human-generated data on travel behavior and well-being. Daynamica can be adapted for a variety of uses and can be personalized to deliver adaptive behavioral integrations.

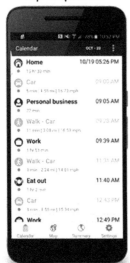

a. **Calendar View** displays auto-detected activity-trip sequence.

b. **Survey View** asks for additional self-reported activity/trip details.

c. **Map View** displays spatial details of activities and trips.

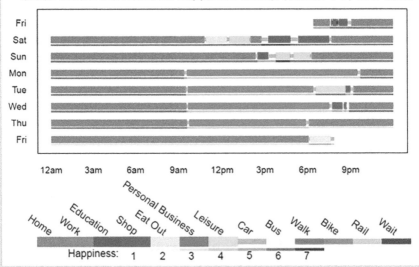

d. **Time Use and Well-Being Data generated by the app.**
Wider bars indicate time use associated with activities and trips.
Green thinner bars indicate happiness ratings (scaled 1-7) over time.

Reprinted with permission from © Daynamica.

The novelty of the Daynamica app is twofold:

1. The use of spatiotemporal machine learning algorithms to reconstruct daily time use (activity) and spatial movement (trip) episodes from mobile sensing data in real time.
2. The use of a well-designed user interface to acquire additional self-reported data on the autoconstructed daily activity and trip episodes such as emotional well-being.

The technology, initially called SmarTrAC [1] and later named Daynamica, was granted a U.S. Patent titled "Travel and Activity Capturing" in September 2017 [2]. A year later, Fan and her collaborators founded Daynamica Inc. with the help from the University of Minnesota Venture Center. The Daynamica app is currently available in both the Apple App Store [3] and the Google Play Store [4]. The app has been deployed in a wide range of research studies across the United States to collect travel behavior and well-being data [5, 9, 10, 11, 12, 13]. Recently, Daynamica Inc. has expanded the capabilities of the app by integrating physical activity and biometric data from wearable devices (including the Modus StepWatch™ and Empatica E4 wristband) [6] and developing the capability to deliver context-sensitive behavior interventions within the app to promote more sustainable and happier travel behavior [7]. The Daynamica team has also developed a HIPAA-compliant version of the app that is currently being trialed in two medical research studies at the University of Minnesota. Figure 2.1 shows the current app interface and the activity- and trip-level time use and well-being data generated by the app.

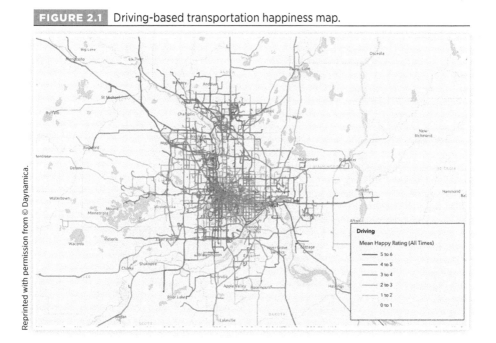

FIGURE 2.1 Driving-based transportation happiness map.

Results

One example of Daynamica in application is the Minneapolis-St. Paul Transportation Happiness Map [8]. This map illustrates spatiotemporal differences in travelers' happiness ratings on the streets and roads in the Minneapolis-St. Paul metropolitan region. The data powering the map was collected using the Daynamica app. In the first map (Figure 2.1), we see driving happiness at all times in the recorded period. Happiness is marked in the darker colors. The data shows that commuters who take the scenic West River Parkway in Minneapolis during weekday morning rush hours—whether by car, bike, or foot—are among the happiest in the Twin Cities area.

When comparing between modes, biking is generally happier than driving and other modes. The biking-based transportation happiness map (Figure 2.2) has a much smaller coverage area than the driving-based map because the majority of trips in the dataset are driving trips.

Similarly, the bus-based transportation happiness map (Figure 2.3) has a small coverage area due to the limited bus trips in the dataset. Nonetheless, the map illustrates spatiotemporal differences in travelers' happiness ratings on various bus routes. For transit planners who are interested in improving people's transit experiences, the map provides important insights on specific bus routes and road segments that are in need of closer investigation for future improvements.

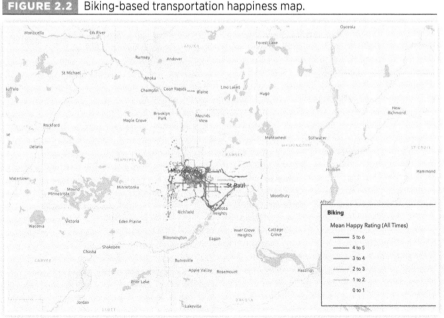

FIGURE 2.2 Biking-based transportation happiness map.

FIGURE 2.3 Bus-based transportation happiness map.

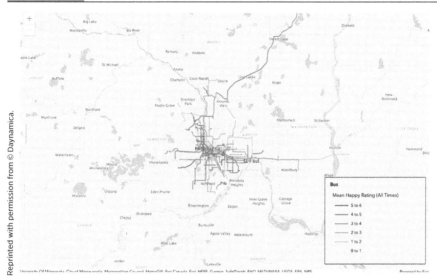

To allow users to explore the mode-specific and time-varying nature of transportation happiness, we not only calculated the mean and median of each emotion variable of all trips matched to a single segment but also calculated the means and medians among trips of different time bucket categories and different travel mode categories. In the end, given the mode/time/emotion/statistic variations, the Minneapolis–St. Paul Transportation Happiness Map allows a total of 1,120 combinations of mapping options (10 mode options * seven-time options * eight emotion options * two statistic options). These options are listed in Table 2.1. The net effect and peak effect values were calculated as below:

- Net effect = Mean (happy, meaningful) − mean (sad, pain, tired, stress).
- Peak effect = Max (happy, meaningful) − max (sad, pain, tired, stress).

TABLE 2.1 Available mapping options at the Minneapolis–St. Paul Transportation Happiness Map.

Mode options (10)	Time buckets (7)	Emotion ratings (8)	Statistics (2)
All modes	All times	Happy	Mean
Driving	Weekday a.m. rush	Meaningful	Median
Bus	Weekday p.m. rush	Sad	
Walking	Weekday a.m. + p.m. rush	Pain	
Biking	Weekday non-rush	Tired	
Rail	Weekday all times	Stress	
In vehicle	Weekend all times	Net effect	
Driving + in vehicle		Peak effect	
Walking + biking			
Bus + rail			

Daynamica has been deployed in U.S. cities including Minneapolis, MN; Washington, DC; and El Paso, TX, and has shown that across modes biking was the happiest and bus transit the least happy. Both findings validated prior studies.

About the Team

Yingling Fan is Professor of Urban and Regional Planning at the University of Minnesota and CEO at Daynamica Inc. Her research examines the impacts of urban environments on human behavior and well-being and identifies built environment solutions for positive social equity and human health outcomes in cities and metropolitan regions. Her work has received more than U.S. $18 million in extramural research funding, including significant contributions from the USDOT, the U.S. Department of Housing and Urban Development (HUD), the U.S. National Science Foundation, and the National Institutes of Health. She was the lead investigator of the five research projects between 2011 and 2018 that developed the smartphone-human hybrid intelligence system to measure and shape human behavior and well-being (namely, the Daynamica system). Using the Daynamica system, she has pioneered the transportation happiness concept and developed systems to measure, analyze, and visualize individual- and population-level travel behaviors and their associated emotional experiences. Dr. Fan and her collaborators founded Daynamica Inc. and received the worldwide exclusive license to commercialize the Daynamica system (U.S. Patent: #9763055B2) in 2018. She has served as CEO since the company's establishment. Since 2018, the company's mobile app development efforts have expanded into diverse market sectors including transportation planning, urban design, public health, and remote patient monitoring. Dr. Fan has a PhD in city and regional planning from the University of North Carolina at Chapel Hill and a bachelor's degree in transportation engineering from Southeast University, Nanjing, China.

References

1. Fan, Y., Wolfson, J., Adomavicius, G., Vardhan Das, K. et al., "SmarTrAC: A Smartphone Solution for Context-Aware Travel and Activity Capturing," Center for Transportation Studies, University of Minnesota, 2015, accessed November 1, 2020, http://www.cts.umn.edu/Publications/ResearchReports/pdfdownload.pl?id=2588.
2. Fan, Y., Wolfson, J., and Adomavicius, G., Travel and activity capturing. U.S. Patent No. 9,763,055B2, U.S. Patent and Trademark Office, Washington, DC, 2017, https://patents.google.com/patent/US9763055B2/en.
3. Daynamica Inc., "Daynamica 4+: Making Sense of the Day," 2020, https://apps.apple.com/us/app/daynamica/id1507309594.
4. Daynamica Inc., "Daynamica—Making Sense of the Day," 2020, https://play.google.com/store/apps/details?id=com.daynamica.smartrac.
5. Fan, Y., Brown, R., Das, K., and Wolfson, J., "Understanding Trip Happiness Using Smartphone-Based Data: The Effects of Trip- and Person-Level Characteristics," *Findings* (2019): 4-13, doi:https://doi.org/10.32866/7124.
6. Daynamica, Inc., "Our First Wearable Integration," 2020c, https://daynamica.com/2020/03/26/our-first-wearable-integration/.

7. Fan, Y., Becker, A., Ryan, G., Wolfson, J. et al., "Smartphone-Based Interventions for Sustainable Travel Behavior: The University of Minnesota Parking Contract Holder Study," University of Minnesota Center for Transportation Studies Report 20-13, 2020a, https://hdl.handle.net/11299/218025.

8. Fan, Y., Ormsby, T., Wiringa, P., Liao, C.-F. et al., "Visualizing Transportation Happiness in the Minneapolis-St. Paul Region [Machine-Readable Database]," 2020maps, University of Minnesota, Minneapolis, 2020b, https://maps.umn.edu/transportation-happiness/.

9. Le, H.T., Buehler, R., Fan, Y., and Hankey, S., "Expanding the Positive Utility of Travel through Weeklong Tracking: Within-Person and Multi-Environment Variability of Ideal Travel Time," *Journal of Transport Geography* 84 (2020): 102679.

10. Glasgow, T.E., Le, H.T., Geller, E.S., Fan, Y. et al., "How Transport Modes, the Built and Natural Environments, and Activities Are Associated with Mood: A GPS Smartphone App Study," *Journal of Environmental Psychology* 66 (2019): 101345, doi:10.1016/j.jenvp.2019.101345.

11. Ambrose, G., Das, K., Fan, Y., and Ramaswami, A., "Is Gardening Associated with Greater Happiness of Urban Residents? A Multi-Activity, Dynamic Assessment in the Twin-Cities Region, USA," *Landscape and Urban Planning* 198 (2020): 103776.

12. Lal, R.M., Das, K., Fan, Y., Barkjohn, K.K. et al., "Connecting Air Quality with Emotional Well-Being and Neighborhood Infrastructure in a US City," *Environmental Health Insights* 14 (2020): 1178630220915488.

13. Brown, R., "A Statistical Framework for Harnessing Human Activity Data to Understand Behavior, Health, and Well-Being," PhD dissertation, 2020, https://conservancy.umn.edu/handle/11299/213122.

14. Fan, Y., "NYC DOT's Annual Mobility Management Conference," 2016, https://www1.nyc.gov/html/dot/downloads/pdf/mm-conference-program.pdf (keynote speaker).

From Revolutions per Minute to Experiences per Mile

Karen M. Piurkowski, Global Automotive Marketing Director, Co-founder of the Experiences per Mile Advisory Council, Host of the *Experiences per Mile* podcast, Moderator for HARMAN Experiences per Mile webinars and virtual events

Background

The metrics consumers use to measure a car's worth are shifting as they explore new ways to connect with their vehicles in the new era of mobility. To complement their increasingly connected lives, consumers are looking beyond horsepower and brand names and are instead paying more attention to how they can remain connected even when on the go. As consumers continue to explore new ways to connect to their world, both inside and outside their vehicles, it is no surprise that the way consumers measure a car's worth is moving away from dynamic performance alone—a shift from revolutions per minute (RPM) to experiences per mile (EPM). To keep up with the interests and changing lifestyle demands of consumers, automotive companies must find new ways to adapt or risk being left behind.

In anticipation of this monumental shift impacting the industry, HARMAN and SBD Automotive teamed up to create the EPM Advisory Council in October 2019.

HARMAN designs and engineers connected products and solutions for automakers, consumers, and enterprises worldwide, including connected car systems, audio and visual products, enterprise automation solutions, and services supporting the Internet of Things (IoT). More than 50 million automobiles on the road today are equipped with HARMAN audio and connected car systems, and the design process puts consumer understanding—from the car to the living room and beyond—at the core of everything it does. With this expertise, HARMAN, along with automotive technology and security consultants SBD Automotive, sets out to address this industry change by recruiting experts from across the mobility sector to join the EPM Advisory Council.

Goals and Objectives

The goal of the EPM Advisory Council is to encourage collaboration among an exclusive group of mobility executives, analysts, and industry insiders regarding the changing value chains in automotive being led by the connected movement (Figure 2.4). The EPM Advisory Council members are playing a special role in writing the next chapter in automotive history by solving the complex issues facing the industry both today and tomorrow. Specific goals of the EPM Advisory Council include:

- Form cocreated EPM Thought Leadership reports.
- Evangelize the EPM messaging in the industry.
- Create an industry-wide metric to measure EPM.
- Network and collaborate with other members from diverse industries.

Ultimately, the EPM Advisory Council seeks to challenge automotive and nonautomotive companies to rethink the optimum connected vehicle experience, meet the needs of the demanding personalized experiences that today's consumers seek, and make significant changes in the evolution of the mobility industry.

Approach and Challenges

When it was formed, the EPM Advisory Council [SJ2] sought to attract members from the most innovative companies in the mobility ecosystem. Qualified participants were selected from OEMs (original equipment manufacturers), automakers, automotive parts suppliers, and third-party suppliers as well as technology organizations, application providers, financial services, and more. This curated membership would ensure a diverse consensus on how to collectively move the industry forward.

Council members serve as active participants in conversations that directly impact the direction of our industry. Members are expected to attend quarterly council meetings either in person or virtually, actively participate in the discussions during each meeting, and participate in ongoing communication between engagements to further facilitate the evolution of in-vehicle experiences. Additionally, the council is grouped into various committees in which executives with various backgrounds can work together to solve specific issues. Those specialized committees include the Metrics Committee, Technical Subcommittee, and Membership Committee.

FIGURE 2.4 Experiences Per Mile 2030: Ensuring the Next Decade of Mobility Transformation Puts the Consumer First and Foremost.

A Clear Roadmap for the Next Decade of Mobility Success

Achieving all of the above may be the toughest transformation yet for an ecosystem that is by now well-accustomed to disruption. It won't happen at once—In fact it needs to happen in a sequence that helps stakeholders realign in a sustainable manner. Below is a roadmap for how we see this transition occurring:

Today

Experimentation and Learning
Passengers and companies are taking risks and learning new ways of "being moved". Customers are experiencing new forms of mobility, paying, accessing services, delivery, and companies are investing billions in testing new business models.

2021-2023

Wisdom Built on Deep Insights
Insights drive more to an ecosystem mindset and accelerate collaboration. Data and insights provide new players and existing investors to think in new ways of enhancing the customer experience. Companies start working together to enable the consumer to make optimal choices about their mobility experience.

2024-2027

The Experience Economy
The car is a favorite place. Consumers see the vehicle as a space to be more entertained, more productive, more...everything. The growing maturity of higher-levels of autonomy helps consumers regain much-valued time, and their mobility solutions help them make the most of that time through improved productivity, social interactions or healthy living.

2028-2030

Meshed Ecosystem of Mobility and Living
Break the mold of the form factor—the "vehicle" is the door. It brings "destinations" to you, rather than taking you from point A to point B. In this era, the home is as much a part of the mobility experience as the car. This is a time when you can delegate to the "vehicle," where it can pay, transact and complete tasks for you. Mobility providers build integrated experiences into all products in a consumer's life: home, car and devices. Consumers evolve from "Owning a Premium Car" to "Living a Premium Life."

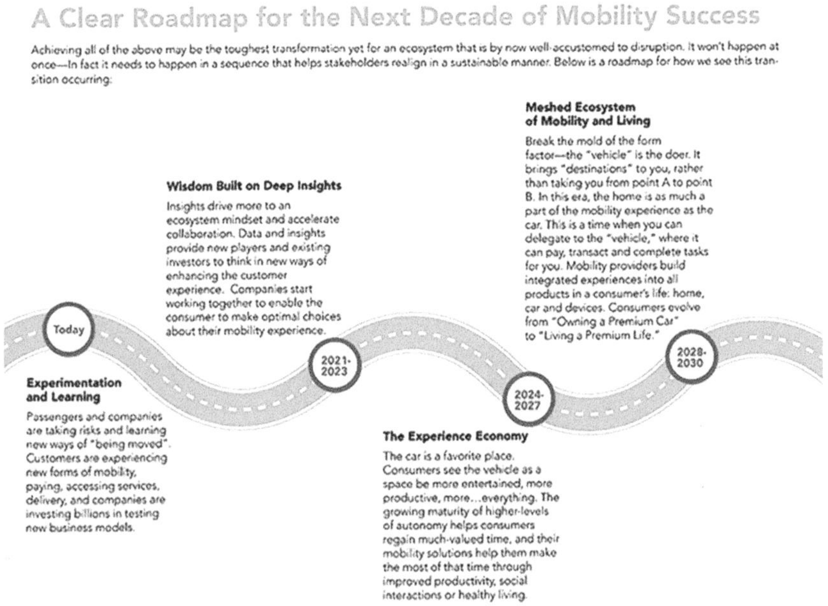

Reprinted with permission from © 2021 HARMAN INTERNATIONAL INDUSTRIES, INCORPORATED. All rights reserved. Features, specifications and appearance are subject to change without notice. All company and product names used herein may be trademarks of their respective owners.

The EPM Advisory Council's activity centers on collaborative meetings with interactive roundtable discussions that encourage an open dialogue about what could and will be accomplished by collaborating across the industry. The number one goal of this program is thought leadership and enabling a consumer-centric vision, which gives participants a more collaborative and informed understanding of what defines mobility, where it is heading, and how to capitalize on this paradigm shift.

Results

The EPM Advisory Council includes members from the most innovative companies in the mobility ecosystem. Today there are 22 companies and 33 members involved in the EPM Advisory Council offering a global perspective on mobility and connectivity solutions. Some of the most prominent names in the automotive and connected ecosystem space take part in the council, including:

- AccuWeather Amazon Web Services, Cisco, Cox Automotive, Darktrace, Ford, GM, HARMAN, HERE Technologies, Hyundai, McLaren Automotive, Nissan, Otonomo, Polaris, Salesforce, Samsung Electronics, SAP, SBD Automotive, Spotify, Stellantis, TomTom, and Zoox.

Recent outcomes of the EPM Advisory Council are the formation of two industry reports. The first report, "Experiences Per Mile 2030: Ensuring the Next Decade of Mobility Transformation Puts the Consumer First and Foremost," addresses the massive mobility transformation currently underway and diagnoses why consumers are not getting the most out of today's mobility model.

The most recent report, "Experiences Per Mile: Charting an Ambitious Course to Measure Mobility Experience for the First Time," identifies how current research methodologies have fallen short of measuring the holistic mobility experience, what constitutes a useful and actionable metric, and how the EPM Advisory Council is leading the industry toward a new way forward.

The EPM Advisory Council is currently working on a proof of concept that will be used as the base for developing an industry-wide metric to measure EPM. To create this new measurement, a Metrics Committee was formed in September 2020. Biweekly meetings have been held to define the scope of work. Since then, the committee has developed a description of the outcomes expected from the initial research and has spoken to 14 research providers to select the company best suited for the research. Research will begin soon, and results will be published at the end of the third-quarter of 2021.

The EPM Advisory Council is continuing its work and will be celebrating its two-year anniversary in October 2021.

About the Team

Karen M. Piurkowski has more than 20 years of experience in business development, managing product lifecycles, brand building, strategy formulation, business operations, product launches, market research, and team leadership for technology, automotive, manufacturing, energy, retail, health care, and financial services industries. A decisive leader, she is an expert in implementing sound business practices to achieve turnaround growth, positioning organizations for long-term profitability, and creating ethical sales models using her vast knowledge of both the market and the capabilities of the product.

Currently, the Global Automotive Marketing Director at HARMAN International, Piurkowski is responsible for all aspects of marketing and communications. She is Co-founder of the EPM Advisory Council, Host of the *Experiences Per Mile* podcast, and Moderator for HARMAN EPM webinars and virtual events. She developed her strategic marketing expertise while conceptualizing, planning, and executing programs for six major corporations, including Harman International, OpenText (Covisint Corporation), Compuware, Equifax Marketing Services, MSX International, and IHS Automotive driven by Polk. Piurkowski has been an active member of the Michigan Council for Women in Technology since 2008. She has also been an active member of the Automotive Industry Action Group for nine years, where she served on the Supply Chain Steering Committee and won an Outstanding Achievement Award in 2010. She obtained a position on the Ferris State University Marketing Advisory Board in 1994 where she assisted in determining the curriculum for the students seeking marketing undergraduate degrees. She served as an MSX International and IBM Governance Board Chairperson in 2002.

Other female members of the EPM Advisory Council leading the way in mobility include:

- May Russell, Emerging Technologies Executive, Chief Information Officer (CIO), Ford Commercial Solutions, Ford Motor Company.
- Charity Rumery, Vice President (VP), Americas Automotive and Industrial, HERE Technologies.
- Bita Sistani, Director, Automotive Marketing, Samsung Electronics.
- Lea Malloy, Assistant Vice President, Emerging Technology, Cox Automotive.

3

Funding

Rawpixel.com/Shutterstock.com.

The transportation industry is experiencing an enormous disruption that can result in wonderfully positive changes to our lives. We invest in startups that are making our transportation safer through technology and autonomy, cleaner through electric vehicles, and more accessible for everyone through micromobility, car sharing, and mass transit.

—Kathryn Schox, General Partner, Trucks Venture Capital

Just like any other business endeavor, mobility companies rely on the funding they receive and the support of their investors. Funding is crucial to get a business, a project, or research up and running. Funding ensures that innovative ideas can become realities.

But the reality is that finding and securing funding is difficult—especially for women.

A recent analysis from Pitchbook found that in Q3 2020,[1] venture funding for female startup founders hit its lowest quarterly total in three years—the lowest figure since 2017. The funding challenges experienced by female founders this past year are especially troubling, considering that it comes on the heels of a record funding cycle the year prior. In the first quarter of 2019, Venture Capital (VC) funding hit a decade peak of $950 million, and the year ended with a total of $3.35 billion invested.

The dip in funding can be partially attributed to the COVID-19 pandemic, as figures demonstrate all funding data was low in 2020; however, female-led companies received disproportionately lower funding when compared to their male-led counterparts.

The gender gap in investing is a poor business deal for more than just the obvious—that diversity can be rewarding to people, but it is also good business. When it comes to VC, greater diversity yields higher returns. According to a Boston Consulting Group (BCG) study, women-founded startups are a better investment.[2] The BCG found that for every dollar of funding, these startups generated 78 cents, whereas male-founded startups generated only 31 cents.

A lack of female-led funds is one of many reasons why the funding gap for women continues to be wide. Fewer than 10% of decision-makers at U.S. VC firms are women. This means that even though dozens of firms have made concerted efforts to diversify their ranks, only 105 investors out of roughly 1,088 were female. This truly complicates things.

In an industry dominated by men, such as mobility, it is even more difficult for women-led companies and projects to access capital and/or obtain the capital that they need. Automotive is a boys' club, technology is a boys' club, and so is funding.

To counteract the lack of female funding, there has been a rise in female funders and investors who are dedicated to ensuring the diverse support of entrepreneurs and innovators. Additionally, in 2019, Ford Motor Company launched a grant program called SHE-MOVES along with the Ford Fund that is dedicated to empowering women to make a positive impact on their communities through mobility services, particularly related to electrification. With an aim of empowering women in the United States as well as South Africa, SHE-MOVES awards $50,000 per project.

In this book's foreword, you read a piece penned by Governor Gretchen Whitmer of the State of Michigan. She shared the commitment and support for mobility innovation and the industry as a whole that she and her administration provide at a state level. As part of this commitment, the state has spent the last few years funding transportation projects through grants, specifically projects that push the envelope on how we view mobility today. This funding project is one of the featured case studies

---

[1] https://pitchbook.com/news/articles/vc-funding-female-founders-drops-low
[2] https://www.bcg.com/en-us/publications/2018/why-women-owned-startups-are-better-bet

that you will read in this chapter. Along with it, there are examples of female-led projects that either need funding or have successfully secured it. Each story highlights just how vital the cash flow can be and how it can make a difference even in the most remote communities.

An Economic Development Approach to Funding Mobility Projects, 2018–2020

Amanda Roraff, Co-founder, AV Mobility Corridor Team, Head of Engagement, Ford Motor Company (formerly Managing Director, PlanetM, Michigan Economic Development Corporation)

Kathryn Snorrason, Managing Director, Michigan Office of Future Mobility and Electrification (formerly Strategic Accounts Director, PlanetM, Michigan Economic Development Corporation)

Kate Partington, Program Specialist, Michigan Office of Future Mobility and Electrification (formerly Program Analyst, PlanetM, Michigan Economic Development Corporation)
Michigan

Background

Since the first Model T rolled off an assembly line, Michigan has been the global leader in the mobility sector. Maintaining this leadership is imperative to the continued prosperity of the state and its residents.

Many of the state's government agencies, universities, accelerators and incubators, startups, investors, municipalities, automakers, and suppliers were focused on the future of mobility but were not necessarily working together in a cohesive way that would most effectively and efficiently advance Michigan into the global leadership position that we hold today. In 2016, the PlanetM brand was born to further accelerate the future of mobility in Michigan and secure the state's global leadership position.

PlanetM, part of the state's economic development arm—Michigan Economic Development Corporation (MEDC)—was the first-ever government entity focused on creating new opportunities to bring together key stakeholders to accelerate the future of mobility.

With full understanding that the future of mobility will not be solved by any one entity or one state, and operating in a competitive market of Michigan, PlanetM operated under the belief that through collaboration Michigan could be a key player in accelerating that future by growing an ecosystem dedicated to safer, more equitable, and environmentally conscious transportation.

For mobility startups specifically, one of the key barriers to their growth was the need for funding to deploy their next-generation transportation solutions in a real-world environment. In response, PlanetM launched the PlanetM Mobility Grant program in 2018 and, over time, expanded the grant program to help companies access several of Michigan's AV and mobility test sites.

Goals and Objectives

The PlanetM Mobility Grant program was created to help mobility startups and corporations to deploy their technologies in Michigan or prove out their technology at Michigan's state-of-the-art testing facilities. The main goal of the grants was to fund innovative technology activations with the intention to scale these solutions in ways that could impact Michiganders statewide, not just where the pilot project was launched.

The secondary goal supported economic development for the state. PlanetM was committed, through this grant program and other initiatives, to demystify government, cut through red tape, connect people with decision-makers, and develop funding opportunities to deploy real-world solutions.

Approach and Challenges

The PlanetM Mobility Grants were launched in 2018 and were separated into two different pathways: pilots and testing.

The Pilot Grants support global mobility companies in deploying their technologies throughout Michigan in partnership with Michigan communities. Companies awarded with pilot grant funds conduct real-world testing and are ultimately focused on making it easier, safer, and more affordable for people and goods to move around.

The Testing Grants were meant to accelerate innovation by providing mobility companies the opportunity to access Michigan's advanced, state-of-the-art testing facilities including Mcity in Ann Arbor, American Center for Mobility in Ypsilanti, Great Lakes Research Center and Keweenaw Research Center in Michigan's Upper Peninsula, Kettering University's GM Mobility Research Center in Flint, and Michigan Unmanned Aerial Systems Consortium in Alpena. Mobility companies were awarded Testing Grant funds to help reduce the costs associated with testing at these facilities and to allow mobility companies to focus their resources on enhancing and proving out their technologies while deepening ties to Michigan's mobility ecosystem.

The PlanetM team sought projects that were not only "first of its kind" technologies but ones that also addressed a mobility pain point in Michigan communities. By taking this approach at funding, PlanetM aimed to position Michigan as a leader in the mobility space, as well as address critical mobility gaps and assist companies in proving out their technologies.

In March 2020, PlanetM transitioned the PlanetM Pilot Grant funds to the COVID-19 Mobility Solutions Grant to provide funding to mobility solutions that address the crucial challenges presented by the spread of COVID-19 within the state of Michigan.

Results

From the time of launch, PlanetM invested $2.4 million in 26 pilots and 27 test grants. Many of these mobility companies have gone on to secure contracts with municipalities and the automotive industry after they were able to test and deploy their solutions in real-world environments.

The most notable grants included:

- Derq was the first company to receive a pilot grant from the program, and since their award, they have established a presence in Michigan, hired employees here, and are now scaling their intersection safety technology across intersections in Southeast Michigan—securing the M1/Woodward project (announced in 2020) and partnership with Motional with a pilot in Vegas (announced 2021).

- Airspace Link received a $55,000 grant to create a traffic management platform for drones in the state. Since the project was successfully completed, they have scaled their technology throughout the United States and have hired 10 high-wage Michigan-based employees and received more than $2 million in investment.

- Pratt Miller received $50,000 to deploy an autonomous cleaning robot in the Grand Rapids Airport to aid in COVID-19 mitigation. They have since expanded their robots into the Detroit Metropolitan Airport and LaGuardia Airport in New York.

- Propelmee is currently working on setting up a team in Michigan. The company received a $100,000 grant from PlanetM to map all 7,000 miles of major roadways in Michigan. Now Michigan is the only state in the United States to have all of their major highways mapped for AVs.

Next Steps

In February 2020, Governor Whitmer created the Office of Future Mobility and Electrification (OFME) and the Council on Future Mobility and Electrification. The OFME was created to further the state mobility work led by PlanetM and to coordinate with the other state agencies that have a hand in mobility work, including MEDC, MDOT, Labor and Economic Opportunity, and EGLE (Environment, Great Lakes, and Energy). The OFME continues to work across state government, academia, and private industry to enhance Michigan's mobility ecosystem, including developing dynamic mobility and electrification policies and supporting the startup and scale-up of emerging technologies and businesses.

In early 2021, the PlanetM brand was officially sunset, and the former PlanetM team continues to do mobility programming work under the OFME. Based on the demand and positive results from the PlanetM Mobility Grant, the OFME aims to continue to fund innovative mobility pilots and testing across the state.

About the Team

Leading the team from Michigan's PlanetM were Amanda Roraff, who helped to evolve the PlanetM brand into an operating platform to drive innovation in the state, bring stakeholders together to accelerate the mobility industry, and drive economic development before taking on a new role with Ford Motor Company; Kathryn Snorrason as the Strategic Accounts Director, who managed the final round of the mobility grant program in 2020, including the transition to a focus on COVID-19 mobility challenges; and Kate Partington, who, as a Program Analyst, led operations for the grand program including day-to-day communications to applicants and tracking pilot and testing awards.

Amanda Roraff is Co-founder, AV Mobility Corridor Team and Head of Engagement at Ford Motor Company. In her role, Roraff is focused on developing the future of roads to accelerate the adoption of AVs and expand access to transportation. Previously, Roraff was the Managing Director of PlanetM, the State of Michigan's business development program focused on accelerating the global mobility industry. Roraff was also previously the vice president of marketing and communications at NextEnergy, where she led various efforts focused on industry and venture development and helped drive tens of millions of dollars in new investment in supplier diversification efforts. Roraff has served on various boards and councils focused on accelerating the mobility industry. Since 2011, Amanda has been a board member for the nonprofit organizations Our Global Kids, which supports youth educational programs, and the World Affairs Council of Detroit, which provides a forum for global issues.

Kathryn Snorrason is a Managing Director for the OFME for the state of Michigan. Snorrason is focused on growing Michigan's mobility and electrification ecosystem and creating safer, more equitable, and environmentally conscious transportation solutions for Michigan residents. Prior to joining the OFME, Snorrason was a consultant at Deloitte Consulting and a manager in the economic development space with World Business Chicago. Snorrason graduated from the University of Michigan Stephen M. Ross School of Business with a Bachelor of Business Administration and a minor in community action and social change.

Kate Partington is Program Specialist at the OFME for the State of Michigan. Prior to the OFME's creation in 2020, Partington worked on both PlanetM (mobility) and Pure Michigan Business Connect (supply chain) programs at MEDC. Prior to her time at MEDC, Partington worked for Senator Debbie Stabenow in her Washington, DC office and held internships in the Planning and Development Department and Department of Neighborhoods at the City of Detroit.

Regional Transit Authority of Southeast Michigan Ballot Initiative

Tiffany Gunter, Regional Transit Authority of Southeast Michigan, Interim CEO
Southeast Michigan

Background

After 23 previously unsuccessful attempts to establish an effective regional transit authority, and as a response to Michigan's inadequate public transit design, funding, investment, and adoption, the Regional Transit Authority (RTA) of Southeast Michigan was established in 2012 by Public Act No. 387, the Regional Transit Authority Act (Figure 3.1). This legislation intended to assist regional transit authorities in providing regional transportation in Southeast Michigan.[3] It has a 10-member board representing the area of Washtenaw, Wayne, Oakland, and Macomb counties. The RTA's mission states: "Our mission is to manage and secure transportation resources that significantly enhance mobility options, to improve quality of life for the residents and to increase economic viability for the region."

Goals and Objectives

The RTA was tasked with the following objectives:

- Optimize existing services of five public transit providers:
  - Detroit Department of Transportation (DDOT).
  - Suburban Mobility Authority for Regional Transportation (SMART).
  - Ann Arbor Area Transportation Authority (AAATA).
  - Detroit Transportation Commission: Detroit People Mover.
  - Q-Line.
- Develop a four-county master transportation plan.
- Secure funding for implementation of the master transportation plan for 20 years.

FIGURE 3.1 Rendering of BRT on Michigan Avenue.

---

[3] http://www.legislature.mi.gov/(S(s4h1svfddsefus3evwjryrmd))/mileg.aspx?page=getobject&objectname=mcl-Act-387-of-2012#:~:text=AN%20ACT%20to%20provide%20for,bonds%20and%20notes%3B%20to%20collect

The leading objective was to optimize existing services and secure funding for higher-level transit services such as Bus Rapid Transit (BRT) and Commuter Rail. The transportation master plan served as the "why" and the "how" to the equation. The funding was created by an increase to the gas tax, and under the legislation, it would support a 110-mile BRT system that would connect Southeast Michigan (Figure 3.2).

The RTA transportation master plan included lines on Gratiot Avenue, Michigan Avenue, Washtenaw Avenue, and Woodward Avenue: all mail arterial corridors that run through the county lines. The BRT would offer a limited number of stops, permanent stations, and enough service frequency to provide reliable and fast trips for riders. Though the BRT was the highlight of the RTA transportation master plan, it also included the following elements (Figure 3.3):

- Direct express service to Detroit Metropolitan Airport.
- Commuter express routes.
- Paratransit.
- On-demand transportation services.
- Commuter rail from Detroit to Ann Arbor.
- Local bus service expansions.

FIGURE 3.2 RTA Rapid Transit Line, proposed in Southeast Michigan.

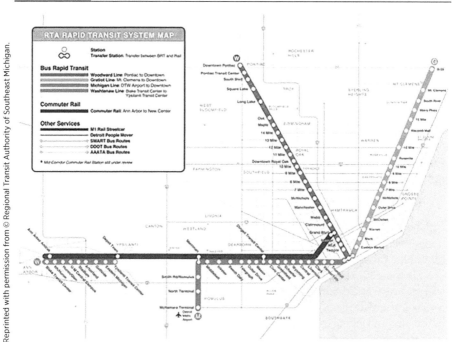

FIGURE 3.3 RTA Proposed Master Plan throughout Southeast Michigan.

Approach and Challenges

The RTA, working with federal, state, county, and local leaders, pursued a ballot initiative in November 2016 that supported the first fiscally constrained transit master plan for Southeast Michigan. Tiffany Gunter was uniquely positioned to lead this effort and was tasked to help manage oppositional attitudes toward transit.

The largest challenge involved was the anti-transit attitude maintained by many southeast Michigan residents, which could result from a lack of education on the benefits of public transportation from both a personal and regional perspective. Personal vehicle ownership in the counties surrounding the Motor City is a legacy perspective that has generational implications, a deep sense of pride in the automotive industry, and a shared lack of interest in any kind of mass transportation.

Many suburban residents did not support the initiative simply because they did not think it would benefit them, or they thought the cost was too high. The proposed millage (the rate at which property taxes are levied on property) was intended to raise $3 billion and would be augmented by nearly $1.7 billion in funds from the state and federal governments. This new tax would have applied to all homeowners in service areas of the AAATA, DDOT. and SMART. The RTA estimated that this tax would have cost the average Southeast Michigan resident $95-$120 annually. Many opponents argued that this tax hike was too high and that these services were already supported by other taxes. There were also claims that the existing transportation network was too outdated to be anchored to by this plan.

Results

On 2016 Michigan election ballots, the measure failed by 1%, just 18,000 votes. The millage was passed in both Wayne and Washtenaw counties, the home counties of Ann Arbor, Detroit, and the International Detroit Metropolitan Airport. Wayne County came in at 360,470 votes in favor (53%) and 323,443 votes against the effort. In Washtenaw County, there were 94,274 votes in favor (56%) and 73,632 votes in opposition. The effort failed in both Macomb and Oakland counties, with 40% and 49.91% in favor of the proposed millage, respectively.

Southeast Michigan's failure to pass a transit millage stands in contrast to most other areas of the United States. In a news release, the American Road and Transportation Builders Association said, "69 percent of the 280 transportation-funding ballot measures up for vote across the nation were approved, with results still pending for seven local areas."

The failed millage leaves the public transportation system of Michigan operating with business as usual. The existing transportation agencies will continue to function under the RTA umbrella, and none of the proposed service expansions and additions moved forward as planned.

Since the 2016 transit referendum's narrow loss, elected, business, and philanthropic leaders have worked diligently with transit advocates, outside experts, and the RTA to review and update the 2016 Regional Master Transit Plan, adopted by the RTA. We began by listening and taking into account the most consistent feedback to the 2016 Master Plan.

About the Team

Tiffany Gunter is Director of External Affairs for the Southeast Michigan Council of Governments (SEMCOG), a membership organization that convenes more than 170 local governments in seven counties in an effort to improve the quality of the region's water, make the transportation system safer and more efficient, revitalize communities, and advance economic development for the seven-county region.

Gunter was the former Deputy City Manager for the City of Birmingham, Michigan. Gunter supported the City Manager, running the day-to-day operations of the municipality, and provided strategic recommendations to improve citizen satisfaction. Gunter's primary responsibilities included evaluation of parking infrastructure and operations where limited capacity, heavy demand, and aging infrastructure were a constant challenge for the city.

Prior to serving with the City of Birmingham, Gunter served as Deputy CEO and Chief Operations Officer (COO) for the RTA of Southeast Michigan. Gunter supported the CEO and the Board of Directors in organizing the effort to improve regional transportation and secure dedicated funding for the RTA. Gunter was responsible for the day-to-day management of the RTA administration and operations. She served the RTA for 4 years and stepped in for 1 year as Interim CEO at the request of the Board of Directors.

Tiffany earned a Master of Public Administration and Bachelor of Business Administration, both from the University of Michigan-Dearborn. She also earned a Certificate in Executive Leadership and Management from the University of Notre Dame.

Prior to joining the RTA, Gunter served as Executive Dean of Inter-Institutional Affairs and Educational Partnerships at the Wayne County Community College District, where she provided support to the district in the development of relationships and partnerships that strengthen institutional programs, resources, and networks. Gunter advocated for strategic partnerships that advance the mission of the district and the Chancellor's strategic priorities.

Gunter has worked in the automotive industry as Material Cost Manager for Chrysler Corporation (Mopar Division) and as Transportation Project Manager with the SEMCOG with a focus on major capital investments and improved specialized transportation.

Prior to her entry into higher education, Gunter spent a decade with the SEMCOG. Tiffany performed a critical role in the development of the RTA and was genuinely dedicated to the work of the RTA. She is passionate about improving public transportation, understands the challenges of the area agencies, and continues to work to identify solutions. She continues to serve in this capacity as a member of the Board of Directors for Rail~Volution, a national board focused on multimodality with improved public transportation as its core mission.

Gunter is a dedicated member of The Links, Inc. where she serves the Detroit Chapter as Treasurer.

Creating Safe Pedestrian and Bike Routes through Tribal Land, Karuk Tribe

Tamy Quigley, Senior Transportation Planner, Caltrans District 2 Active Transportation Manager

Kendee Vance, Associate Transportation Planner, Caltrans District 2 Native American Liaison

Misty Rickwalt, Director of Transportation, The Karuk Tribe
Happy Camp, California

Background

Along State Route (SR) 96, through the Town of Happy Camp, California, is a significantly disadvantaged community with many families relying on walking or riding a bike to school, work, and shopping and to access medical services, rather than using privately owned vehicles. Happy Camp is also home to the Karuk Tribe, which is characterized as a progressive Native American community that has recognized the need for a multimodal transportation infrastructure to service the unique needs of the region.

According to the California Healthy Places Index for the area of Siskiyou County[4] where Happy Camp is located, more than 10% of workers commute to work by public transit, walking, or cycling. Additionally, in a survey of Happy Camp Elementary School parents, more than half of responding families live within 2 miles of the school; therefore, their children could potentially walk or bike to school. However, fewer

---

[4] https://map.healthyplacesindex.org/

than 5% of children were walking to school at the time of the survey. Most parents believe walking is dangerous and indicated they will not let their children walk to school regardless of the student's age. When parents were asked which issues affected their decision to allow or not allow their children to walk or bike to school, the most common answers were speed and amount of traffic along the route, absence of sidewalks or pathways, safety of intersections, unimproved crossings, and weather.

Currently, the town has no pedestrian infrastructure, sidewalks, or ADA-compliant curb ramps, and in most locations, very little shoulder along the roadway, which is often used to illegally park. In addition to safety concerns, Happy Camp has challenges that create the need for residents to walk and/or bike. According to the *Estimating Bicycling and Walking for Planning and Project Development: A Guidebook*[5] communities such as Happy Camp that exhibit each of the below characteristics are more likely to need to walk or bike for their daily trips:

- Households with no vehicles or only one vehicle.
- High proportion of residents without a high school education.
- Higher proportion of households living below the poverty level.

Each of these factors leads to the need and importance of these critical multimodal infrastructure improvements.

The tribe is continually pursuing ways to enhance their communities, increase educational opportunities, support tribal enterprises, improve the safety of travelers on their transportation network, and efficiently program their Tribal Transportation Program (TTP) Planning funds to best meet the needs of the tribe. This is how the Happy Camp Complete Streets Project began.

Goals and Objectives

Happy Camp Complete Streets aims to directly address deficiencies in the active transportation network along the entire downtown corridor by creating connectivity between people and places and providing economic prosperity for the community. Additionally, planning for both future developments, new construction, and improvements to existing transportation facilities remains a main goal of the tribe.

The elements of the plan were to provide a flexible model to increase safe walking and multimodal transit, addressing many pressing public health issues facing children and families today by:

- Developing critical pedestrian and bicycle infrastructure.
- Reducing injuries associated with walking and bicycling.
- Increasing daily physical activity levels to reduce obesity and other health risks.
- Reducing vehicle emissions and improving air quality.
- Improving academic performance among children.
- Increasing neighborhood and social cohesion.

---

[5] https://www.nap.edu/read/22330/chapter/1

The proposed improvements would encourage those parents who are reluctant to let their children bike and walk due to safety concerns to have more confidence in letting their children safely walk and bike to school.

It is important to note for indigenous communities that are often geographically, sociopolitically, and economically isolated, as well as culturally, spiritually, and economically dependent on the lands and waters in their region; the health impacts of climate change can quickly become amplified making the need for preservation of clean air and reducing pollution a top priority.

The Happy Camp Complete Streets Project will also directly contribute to achieving the following goal identified in the Karuk Tribe's Climate Adaptation Plan—to improve the mobility, connectivity, accessibility, health, environment, and safety concerns felt by the community.

Approach and Challenges

The Karuk Tribe and the community of Happy Camp began identifying deficiencies and unmet needs of their community through a robust and comprehensive active transportation planning effort specifically to address the lack of safe and convenient ways to connect people to places by walking and biking. Along with the many programmed tribal and area development plans, it is apparent that the connection of tribal community members to local goods and services (as supported by pedestrian or bicycle travel) requires significant planning, expansion, and coordination. The pedestrian plan must focus on the condition of the existing trails and pathways utilized by the tribal citizens (youth to elders) to access services by foot.

Residents of Happy Camp voiced their concerns about existing conditions and support for improvements by participating in health surveys as well as active transportation surveys regarding walking and biking in the community. Participants stated that they are most discouraged from walking and biking because of the current lack of sidewalk/shoulder, lack of lighting, unsafe intersections, and unsafe drivers. However, more than half responded that they would consider biking or walking more if there were bike lanes, sidewalks, and/or wider shoulders.

Elements of the project were decided in coordination with extensive public and community outreach, active transportation surveys, health surveys, stakeholder engagement, presentations to multiple boards and councils, and a comprehensive safety assessment of the Happy Camp project area. These elements, meant to address safety and infrastructure issues, include:

- New 4,160 linear feet of sidewalk and 27 ADA ramps.
- A total of 2,700 linear feet of Class II bike lanes and 128 square feet of lane markings.
- One new rectangular rapid flash beacon, six new shortened crossings, four new bulb outs (sidewalk or curb extensions into the parking lane, which significantly improves pedestrian crossings), and six new or improved crosswalks to increase safe crossings.

- A total of 2,957 linear feet of shoulder rehab and/or widening to accommodate multimodal transportation.
- Increased lighting along 2,300 linear feet of roadway.
- New 1,290 linear feet of barrier to keep pedestrians safe.
- Two new speed feedback signs meant to reduce speeds.

Along the planning process, several specific challenges arose including ensuring planning, and engagement was inclusive and adequately represented the community and also ensuring efforts supported a collaborative partnership with the Karuk Tribe that was respectful of its members. Additionally, the timing made things more difficult as the application for the Happy Camp Complete Street project was not only submitted during a global pandemic, but in early September 2020, a fire, the Slater Fire, ignited outside of the Town of Happy Camp and destroyed near half the town.

Results and Next Steps

In early 2021, the Karuk Tribe was awarded the highly competitive Cycle 5 Active Transportation Program grant, totaling nearly $10 million, by the California Transportation Commission to facilitate a road improvement project in partnership with Caltrans. The award will fund the initial phase of the Happy Camp Complete Streets Project, which aims to improve the walkability and general safety of a stretch of Highway 96 that bisects the rural community of Happy Camp.

Construction is scheduled to begin in the summer or fall of 2025. When completed, the project will allow town planners to build and connect more bike and pedestrian paths to the main street, eventually creating a network throughout the area for residents and outdoor recreation tourists.

The Karuk Tribe recognizes that without the planning process stated above they would likely not be in a position to adequately capture the true needs of all community members, which are well demonstrated in the Happy Camp Complete Streets Projects. Because of the invaluable importance of the planning process, the Karuk Tribe intends to continue to pursue TTP funding in order to supplement projects that can include Roadway Safety Audits, Pedestrian and Bicycle Safety Plan, Trails Plan and Design, Highway Safety Manual Study Analysis of Crash Modification Factors, Corridor Safety Plans, Climate Adaptation Planning, and more.

About the Team

This project is a direct result of a community-driven process initiated by the Karuk Tribe and the community members of Happy Camp. From 2011 to 2020, numerous stakeholders have been engaged and will continue to be engaged throughout the life of this project including community members, students, parents, schools, business owners, various Karuk Tribe departments, Siskiyou County (Local Transportation Commission, Public Works, and Department of Health), state agencies (Caltrans and California Highway Patrol), local law enforcement, Happy Camp Emergency Medical Services and Ambulance, Happy Camp Community Group and recreation groups, and residents living along the SR 96 corridor.

The Technical Advisory Committee (TAC) for the Happy Camp Complete Streets Project consists of:

- Caltrans District 2 Staff (Planning, Engineering, Traffic).
- Karuk Tribe Department of Transportation.
- Karuk Tribe Department of Health.
- Siskiyou County Regional Transportation Agency.
- Siskiyou County Public Works Department.
- Siskiyou County Department of Public Health.

The Happy Camp Complete Streets Project also has support from the following organizations: Karuk Tribe, Caltrans District 2, Siskiyou County, Hope for Happy Camp, Klamath National Forest Service, Karuk Tribe Housing Authority, Happy Camp Volunteer Ambulance Service, Siskiyou County Sheriff, and California Highway Patrol.

Coordinating efforts between all partners, the community, and the TAC are three women representing varying aspects of the Happy Camp region: Misty Rickwalt of the Karuk Tribe and Kendee Vance and Tamy Quigley of Caltrans, District 2.

Misty Rickwalt, Director of Transportation for the Karuk Tribe, works to provide safe and reliable transportation facilities for all users with an emphasis on upholding the mission of the Karuk Tribe. Through transportation and mobility programs, Rickwalt aims to promote the general welfare of all Karuk People and establish equality and justice for the tribe, as well as restore and preserve tribal traditions, customs, language, and ancestral rights.

Kendee Vance, Native American Liaison of Caltrans, serves the tribes of District 2 to closely develop, foster, and maintain the government-to-government relationships with all 23 tribes in the region, including the Karuk Tribe. She works to ensure equity, especially in the northern rural tribal lands, and to create partnerships with the tribes that span far beyond transportation and mobility.

Tamy Quigley, Complete Streets and Active Transportation Manager of Caltrans, works to ensure equity through partnerships in planning, developing, and building transportation for all users, all ages, and all abilities. She works to foster complete streets by providing a safe and designated space for all users within many small and underserved rural communities.

Tennessee Corridor Fast-Charging Network

Alexa Voytek, Energy Programs Administrator at the State of Tennessee
Tennessee

Background

Throughout 2018, Alexa Voytek (Energy Programs Administrator for the Tennessee Department of Environment and Conservation's Office of Energy Programs [TDEC OEP]) and other TDEC representatives worked alongside a core team of stakeholders, including the Tennessee Department of Transportation (TDOT), electric utilities (the Tennessee Valley Authority [TVA] and others), local governments, universities, EV manufacturers, businesses, and advocacy groups—to develop a shared vision for electric transportation in the State of Tennessee. Together, these stakeholders comprise Drive Electric Tennessee (DET), whose goal is to increase EV adoption in Tennessee from approximately 11,000 EVs in 2020 to 200,000 vehicles by 2028.

Goals and Objectives

In January 2019, DET released the first edition of its Electric Vehicle Roadmap,[6] which identifies "Opportunity Areas" that will increase EV adoption across multiple Tennessee use cases and sectors. Three Opportunity Area committees have since been formed to address various projects and initiatives highlighted in the Roadmap:

- Charging Infrastructure Availability.
- Policies and Programs.
- Awareness.

Following the release of the Roadmap, DET conducted a Statewide EV Charging Infrastructure Needs Assessment[7] to evaluate the condition of Tennessee's current EV charging infrastructure and to identify charging needs and potential geographic locations to support the adoption of 200,000 EVs in Tennessee by 2028 (Figure 3.4). The Needs Assessment concluded that additional EV charging infrastructure was needed on highway corridors to relieve range anxiety and to connect rural and urban areas; highway corridor charging was identified as the best candidate for public investment, whereas other EV charging use cases (e.g., community charging, workplace charging, residential or multiunit dwelling charging) were identified as good

---

[6] https://www.tn.gov/content/dam/tn/environment/energy/documents/Roadmap%20for%20Electric%20Vehicles%20in%20Tennessee_Report.pdf
[7] https://www.tn.gov/content/dam/tn/environment/energy/documents/DET%20Tennessee%20EVSE%20Needs%20Assessment%20-%20Executive%20Summary.pdf

FIGURE 3.4 Tennessee EV charging opportunities.

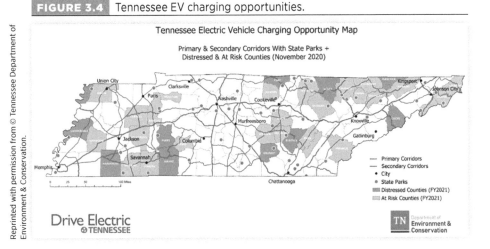

candidates for private or public-private investment based on market attractiveness and anticipated utilization.

Approach and Challenges

To support the goal of establishing a statewide corridor fast-charging network that improves transportation efficiency, reduces vehicle emissions, drives EV adoption, strengthens the resiliency of the transportation sector, and connects both rural and urban areas in Tennessee, Alexa Voytek and other TDEC OEP staff identified key primary (interstates) and secondary (select U.S. and state highways) corridors for EV charging infrastructure development within an EV Charging Infrastructure Opportunity Map.[8] A secondary companion map highlights existing EV fast chargers (direct current fast chargers, or DCFC) along the selected corridors.[9] Each location shown on the existing infrastructure map includes at least one nonproprietary charger, with both J1772 combo combined-charging system (CCS) and CHAdeMO connectors available (these connectors support all EVs except Teslas, which use a proprietary plug type).

The secondary corridors selected for inclusion within the Opportunity Maps were informed by the results of a series of maps generated during the Statewide EV Charging Needs Assessment, which highlighted potential geographic areas for EV charging infrastructure to support the 200,000 EV deployment goal. It also took into consideration Average Annual Daily Traffic data, vehicle-miles traveled (VMT), EV adoption forecasts, and U.S. Census Bureau data. Secondary EV charging infrastructure corridors suggested by the Statewide EV Charging Needs Assessment were then slightly modified to eliminate duplicative, parallel corridors and to prioritize the selection and inclusion of secondary corridors passing through economically distressed and at-risk counties in Tennessee.

---

[8] https://www.tn.gov/content/dam/tn/environment/energy/documents/EVChargingOpportunity-Map_11.20.pdf
[9] https://www.tn.gov/content/dam/tn/environment/energy/documents/EVChargingOpportunityMap_DCFC_11.20.pdf

Because many of Tennessee's State Parks are located in or near rural areas or distressed or at-risk counties, the TDEC OEP also included the locations of Tennessee State Parks on the Opportunity Map to demonstrate how the electrification of the state's corridors can connect EV drivers to the state's natural resources and promote EV tourism in counties that stand to benefit from it most. By prioritizing EV charging infrastructure development on corridors that pass through these counties, the TDEC OEP aims to spur economic activity associated with on-route and/or destination charging in these distressed and rural markets.

The TDEC plans to supplement existing EV charging infrastructure along both primary and secondary corridors and will seek to achieve what the TDEC will call "Fast 50" designation on such corridors by September 30, 2023. To achieve this designation, the maximum driving distance between charging infrastructure locations is 50 miles; each charging location includes two or more fast chargers capable of at least 50 kW concurrent output power each; each fast charger has both CCS and CHAdeMO plugs.

Results

On February 3, 2021, the TDEC and TVA announced a partnership to develop a statewide EV fast-charging network to power the growth of EVs across Tennessee and reduce barriers to transportation electrification. Specifically, the two have signed an agreement[10] to collaborate and fund a network of fast-charging stations every 50 miles along Tennessee's interstates and major highways. This initiative will add approximately 50 new charging locations, tripling the existing fast-charging network. For reference, as of February 2021, there were only 24 fast-charging locations operating in Tennessee that were open to all consumers and supported both charging standards common to EVs.

The TDEC and TVA will leverage various funding sources to support the development of the fast-charging network with an anticipated project cost of $20 million. The TDEC has committed 15%, the maximum allowable, of the state's Volkswagen Diesel Settlement Environmental Mitigation Trust (VW Settlement EMT) allocation[11] to fund light-duty EV charging infrastructures. Approximately $5 million from this fund is expected to be allocated to fast-charging infrastructure along corridors. The TDOT intends to provide the TDEC with an additional $7 million in Federal Highway Infrastructure Program funds to expand fast-charging EV infrastructure along federally designated Alternative Fuel Corridors in Tennessee. The remainder of the project will be funded by the TVA, other program partners, and program participant cost share.

By both spearheading efforts and working with partners on transportation electrification throughout the state, the team at TDEC OEP strives to improve transportation sector efficiency and resiliency as well as to reduce vehicle emissions. Working locally to advance affordable, domestic transportation fuels, energy-efficient

---

[10] https://www.tn.gov/content/dam/tn/environment/energy/documents/TVA-TDEC%20MOA%20-%20EV%20Chargers.pdf

[11] https://www.tn.gov/content/tn/environment/program-areas/energy/state-energy-office--seo-/tennessee-and-the-volkswagen-diesel-settlement.html

mobility systems, and other fuel-saving technologies and practices, the TDEC and its partners can simultaneously reduce energy costs within the transportation sector, promote economic growth, and improve environmental quality.

About the Team

Alexa Voytek is the TDEC OEP Energy Programs Administrator for the office's energy in transportation, communications, and U.S. Department of Energy State Energy Program (U.S. DOE SEP) activities. Voytek serves as Principal Investigator for U.S. DOE SEP in Tennessee, oversees OEP's sustainable transportation and alternative fuels programming, and is the primary point of contact for implementation of the //VW Settlement EMT in Tennessee. Voytek also serves as Coordinator for U.S. DOE's Clean Cities Middle-West Tennessee Clean Fuels Coalition, through which she acts as a technical resource for fleets and individuals evaluating alternative fuels and advanced vehicle technologies. Prior to joining the TDEC, Voytek interned with the United Nations (UN) Division for Sustainable Development and the U.S. Consulate in St. Petersburg, Russia. Most recently, Alexa served as Assistant Account Executive for the public relations firm Ketchum, Inc., where she was assigned to energy- and technology-related projects. Voytek holds a Master of Arts (M.A). from Columbia University in Russian, Eastern European, Balkan, and Eurasian studies and graduated summa cum laude from Duke University, with a B.A. in history and Russian language/area studies.

Voytek is a technical expert on alternative fuels, advanced vehicle technologies, and transportation electrification and is recognized as a national leader and expert on VW Settlement EMT funding, often serving as a subject matter expert for other beneficiaries via the National Association of State Energy Officials (NASEO)/National Association of Clean Air Agencies VW Settlement Working Group. Voytek serves on the leadership team for the NASEO Transportation Committee, through which she supports state energy offices and their partners in transportation-energy business development, environmental protection, and energy reliability. She serves as the Governor's designee to the TVA Regional Energy Resource Council, which provides guidance on how TVA manages its energy resources against competing objectives and values, and is the TDEC representative on the Tennessee Livability Collaborative, a working group of 17 Tennessee State agencies with a mission of improving the prosperity, quality of life, and health of Tennesseans through state government collaboration in the areas of policy, funding, and programming. Voytek serves as Co-chair for the Drive Electric TN Policies and Programs Committee (https://driveelectrictn.org/), a primary point of contact for the Southeast Regional EV Information Exchange, the Tennessee representative on the Climate Registry's Council of Jurisdictions, and a Steering Committee member for TVA's Connected Communities Roadmap development process. Voytek drives a Chevy Volt and is one of 11,034 Tennesseans with a registered EV (as of December 2020).

Voytek serves as Co-Chair for the DET Policies and Programs Committee (https://driveelectrictn.org/) alongside Laurel Creech, Assistant Director for the Division of Sustainability at the Metropolitan Government of Nashville and Davidson County Department of General Services.

4

Infrastructure

Viewfoto studio/Shutterstock.com.

Transportation is such a vital resource that is often overlooked by many of us who can afford to take it for granted. I truly believe that the improvements we make to our transportation infrastructure have the opportunity to bolster the prospects of some of our most vulnerable populations. As leaders in transportation, it is incumbent on us to explore not only how we can use our transportation infrastructure to improve mobility, but also how we're working to enhance communities. Women provide unique and valuable perspectives, and we all have a role to work to craft teams that represent the communities in which we work.

—Erin Slayton, Professional Engineer (PE), DBIA, ENV SP, Transportation Program Management Director, HDR

My daily routine this time last year looked something like this: alarm clock, shower, coffee, walking to the Ferry Street Q-Line stop, logging on to check email while waiting for the streetcar, taking it downtown to my office, working, and then reversing course and riding the streetcar home. Then I would either hang out at home for the night or drive to grab drinks with a colleague or head to an event. If it was a slow night, I would go to the grocery store outside of the city. My days felt busy and back-to-back, and I am certain that until I was in planning school, I hardly considered how the water got to my tap or what was in place underground to allow me to connect to a strong Wi-Fi network while in motion. I would not have thought twice about the condition of the pavement unless it was icy or I hit a pothole. I do not suggest that this is everyone's experience, but I would not be surprised if others could relate to mine. My day-to-day, busy-bee, and productive lifestyle was made possible—and easier—because of the infrastructure that was there to support it. The American infrastructure system includes more than 4.8 million miles of roads, more than 1.2 million miles of plumbing dedicated to drinking water, more than 7,300 power plants, nearly 160,000 miles of high-voltage power lines, and millions of miles of low-voltage power lines and distribution transformers connecting 145 million customers throughout the country.

Advancements in technology have provided the United States advantages (and gaps in access), standardizing and providing options for the way we move access water and other resources, connect, and live. As we advance as a society toward an electrified and more connected future, infrastructure maintenance and improvements will be the test track—adapting to technology, deploying invisible elements of connectivity, and serving as the conduit for a decarbonized transportation and mobility system. Though there are significant critiques of the American infrastructure system, if managed intentionally, it will aid people to move into the next decade more dependent on multimodal mobility options and less dependent on car ownership.

One of the largest challenges the infrastructure system faces is inconsistent and insufficient funding often dictated by the ever-changing partisan agendas associated with both state and federal budgets and priorities. The following case studies offer perspectives into the roads, strategies, networks, and stations that will allow vehicles to participate in a future of electrification and automation.

Infrastructure Connections Making a Difference in Our Communities: I-75 Modernization Project

Barbara Arens, PE, Professional Traffic Operations Engineer (PTOE), Managing Principal, Cincar Consulting Group

Kimberly Webb, Director of Metro Region, Michigan Department of Transportation Oakland County, Michigan

Background

Interstate 75 (I-75) is one of the longest major cross-country, north-south freeways in the United States. Starting in northwest Miami, Florida, and traveling north to Sault Ste. Marie, Michigan, the freeway totals 1,786 miles. The Michigan segment opened in November 1973. Since its opening, the freeway has not received comprehensive corridor improvements but has seen increased traffic as southeast Michigan experienced land-use changes (sprawl) and migration of people through suburban expansion. The Oakland and Wayne Counties stretch of I-75 are critical commuter, commercial, and tourism routes with daily traffic volumes of 103,000–174,000 (pre-COVID-19). The I-75 Modernization Project is roughly an 18-mile project in southern Oakland County, Michigan, from M-102 to south of M-59. It is one of the largest projects in size and scope the MDOT has tackled. The project cost totals $1.7 billion.

I-75 Modernization has been studied like I-94 and other Michigan highways, but due to available funding and community support, the project has moved from planning to engineering to construction over the last 20 years (Figure 4.1). In addition to the High-Occupancy Vehicle (HOV) lane, this project will include Diverging Diamond Interchanges (DDI) and a 4-mile-long drainage tunnel in the southern section. The tunnel is an important improvement to alleviate future flooding in the corridor, as the southern segment is a depressed section of freeway (below grade) that has experienced flooding during heavy rain events in recent years.

FIGURE 4.1 Newly constructed sound wall between I-75 at 10 Mile Road and a residential neighborhood with northbound traffic.

Courtesy of Kristen Shaw.

Projects such as the I-75 Modernization Project are committed to enhancing safety, improving mobility for all existing modes of travel based on each corridor's traveler profile, and meeting the needs of each corridor. The Purpose and Need defined during the Final Environmental Impact Statement of May 2005 stated: "The purpose of the proposed project is to increase the capacity of the transportation infrastructure in the I-75 corridor to meet travel demand for personal mobility and goods movement. Meeting the purpose of the project will improve motorist safety, travel efficiency, and reliability. These are essential both to personal mobility and to the movement of freight."

Goals and Objectives

To achieve a modern design for the corridor, the I-75 Modernization Project adopted an Environmental Impact Statement in 2005 and conducted two engineering reports in 2009–2010. The MDOT worked to find the funding to develop the project into funded segments for construction. The MDOT contracted an Owner's Representative Contract to further assist with the development of the design, assess risks and costs, and utilize alternative delivery methods for the funded construction segments. The 18-mile corridor was split into three segments that, once complete, will improve motorist safety, reliability, efficiency, and environmental sustainability. The completion date for the entire project is estimated to be in late 2023. It will provide the first HOV lane in Michigan, which required legislative changes to allow for HOV. The three segments for the 18-mile corridor are:

- Segment 1: North of Coolidge Highway to north of South Boulevard, Design–Build Contract. Started in August 2016 and was completed in 2017. Reconstructed 3.3 miles, replaced six structures, and modernized the I-75/ Square Lake Road (I-75 Business Loop) interchange.

- Segment 2: North of 13 Mile to north of Coolidge Highway, Design–Build Contract.

    Started in fall 2018 with substantial completion in fall 2020. Project completion is in summer 2021. Reconstructed more than 8.5 miles, replaced 18 structures, and built DDI at Big Beaver and 14 Mile roads.

- Segment 3: North of M-102 (8 Mile Road) to north of 13 Mile Road, Design–Build–Finance–Maintain Contract. Started in winter 2019 with substantial completion expected in fall 2023. This segment will reconstruct 5.5 miles, replace 22 structures (vehicular and pedestrian bridges), remove five bridges, and build a 4-mile-long drainage tunnel in the depressed section of freeway between 8 Mile and 12 Mile roads.

Approach and Challenges

The I-75 Modernization Project plan was developed through a strategic roadmap of traditional steps that lead to successful infrastructure projects. Key elements of the approach include planning, design, and construction phases. The planning stage covers environmental impacts and documents that are required for approvals and ensures that the preliminary design considers historical preservation, air and noise, wetlands, landscaping, key stakeholder, and community engagement. Additionally, this stage includes establishing goals for the aesthetics of the design and the development of the project's purpose and need.

During the design phase, the blueprint for the work is refined, permits are acquired, utility work is established, and stakeholder engagement is continued. The MDOT looked at numerous ways to deliver the I-75 Modernization Project based on the costs and determined how to fund the construction. The MDOT decided to use two different alternative delivery models and to divide the 18-mile corridor into three construction segments; they used design-build and design-build-finance-maintain delivery models. Construction elements have many active elements from utility work to removal and reconstruction of all elements of the freeway and service drives, as well as keeping good communications with key stakeholders and the public through a project website, social media, and public meetings.

HOV lanes, while not new in most states, are new to Michigan, and when the corridor opens these lanes in 2023, they will be the first in Michigan. There is an educational piece to communicate to the public on their purpose and use.

The interesting part of infrastructure is that it is not just engineering. Education is key when infrastructure is new, repaired, or replaced to make sure that people feel comfortable and safe and have a good understanding of the infrastructure use. So, while engineering is important, communication, education, and collaboration elements are key to the successful delivery of a project and program. During 2020 and 2021, the I-75 Modernization Project pivoted to virtual public engagement in response to the COVID-19 pandemic.

Results

Segment 1 (North of Coolidge Highway to north of South Boulevard) began construction in June of 2016 and was completed in September 2017. Segment 1 work included:

- Reconfigured the I-75 Business Loop (Square Lake Road) interchange.
- Replaced freeway lanes and bridges.
- Connected with Bloomfield Township Safety Paths on Squirrel Road over I-75.
- Utilized a landscape plan to plant trees and shrubs.
- Upgraded and added spaces to the existing Bloomfield Township-Adams Road Carpool Lot.
- Incorporated collaboratively developed community aesthetics.
- Addition of an HOV lane in each direction.

Segment 2 (North of 13 Mile Road to north of Coolidge Highway) started construction in fall 2018 and is to be completed in summer 2021. Segment 2 work includes:

- Reconstruction of freeway pavement, interchanges, and bridges.
- Addition of an HOV lane in each direction by enclosing the median.

- Drainage improvements.
- Aesthetics and noise walls where they meet federal/state criteria.
- Innovative interchange design with a DDI at Big Beaver Road and at 14 Mile Road.

Segment 3 (North of M-102 to north of 13 Mile Road) began in fall 2020 and is expected to be completed by fall 2023. Segment 3 work includes:

- Reconstruction of freeway pavement, interchanges, and bridges.
- Additional inside general-purpose lane in each direction, as the HOV lane transitions back to a general-purpose lane from 12 Mile Road to the south.
- Aesthetics and noise walls where they meet federal/state criteria.
- Drainage improvements with the addition of a deep tunnel from M-102 to 12 Mile Road to prevent future flooding events and to separate stormwater from sanitary water.
- Braided ramp to improve safety at the northbound I-75/I-696 interchange. The traffic from the I-696 on-ramp to northbound I-75 will braid (go over) the northbound I-75 traffic exiting to 11 Mile Road (go under). This will separate these important exit/entrance merges and improve safety.

The drainage tunnel from M-102 to 12 Mile Road is an important enhancement, as this 4-mile section of the project is below grade (depressed freeway) and had been prone to flooding during heavy storm events. The tunnel will be 14.5 feet in diameter and 100 feet underground. To create a connection with the public, the tunnel boring machine (TBM) was given a name through a contest. Historically, TBMs are given women's names. So, as a piece of the community engagement strategy, the public were invited to vote on the name of the TBM that would drill the drainage tunnel, and after the votes were in, the TBM took the name of Eliza Leggett, a suffragist and participant in the Underground Railroad in Oakland County.

Once completed in 2023, this 18-mile improvement to I-75 will enhance safety and mobility for Oakland County residents and all of southeast Michigan.

About the Team

Barbara Arens, PE, PTO, has 35 years of engineering, planning, and consultation experience. Since July 2020, she is a Managing Principal at Cincar Consulting Group. Prior to that she was Senior Vice President and Project Manager at WSP USA working for the MDOT on the Engineering Report and Owner's Representative Contract for the I-75 Modernization Project since 2007.

Arens started her career as a traffic engineer and became a major project manager but was also a business manager for an international architectural-engineering company in both operations and business management. Over the last 20 years, she has held management roles locally, regionally, and nationally and has been responsible for more than 700 employees, 15 states, and 13 offices. As a project manager, she has led early preliminary engineering, traffic engineering, environmental impact statements and reports, planning and operational studies, and program management contracts in transportation infrastructure.

Arens uses her experience and passion for people, quality, and clients to drive her career. She is a Director on the Michigan American Consulting Engineers Council. Arens is on the Women's Transportation Seminar (WTS) International Foundation Board and the WTS Central Region Council, and she formerly served as the President on the Michigan WTS Board. Arens is an Emeritus Member of the Transportation Research Board, Transportation Applications (ADB50) Committee. She is a current member of the Conference of Minority Transportation Officials (COMTO), a Fellow in the Institute of Transportation Engineers (ITE), and a member of Chi Epsilon (Civil Engineering Honor Society). Arens was appointed by the Governor of Michigan to serve on the Mackinac Bridge Authority (2014–2018), is a past Board Chair of the Michigan State University (MSU) Civil Engineering Alumni Advisory Board, and won the COMTO "Women Who Move the Nation Award" in 2018.

Arens graduated from Michigan State University with both a B.S. and a master's degree in Civil Engineering. She is a Professional Engineer (PE) in four states and is a national PTOE.

Kimberly Webb, PE, is currently the Metro Region Engineer for the MDOT, where she is responsible for more than 5,000 lane miles of state trunkline (state highway) in the following Michigan counties: Wayne, Oakland, and Macomb. Webb leads a team of more than 300 employees, has a capital annual program of approximately $1 billion, and an annual maintenance budget of $80 million.

In Webb's 31 years with MDOT, she has developed a strong command of the federal, state, and local programs, and has cultivated relationships that positively impact the operations throughout the region and Intelligent Transportation System (ITS), Transportation Service Centers (TSCs). Her extensive experience working in all the major areas of the department's highway operations business, including assignments as a Director of the Bureau of field services engineer, Southwest Region Engineer, Metro Region Deputy Region Engineer, maintenance engineer, TSC development engineer and project manager, Metro Region Associate for Development, and 9 years as the Taylor TSC Manager, has resulted in tremendous leadership progress for and within MDOT.

Webb's influence and leadership extends well beyond the limits of her job duties, most notably through her involvement with COMTO locally and nationally, as well as her involvement with the Women Transportation Seminar (WTS Board and Committee member for American Association of State Highway Transportation Officials (AASHTO). Webb served as Michigan Chapter President of the COMTO for 6 years. She has served as co-chair for COMTO National Scholars Committee, working with youth interested in transportation careers. Webb has made significant contributions to numerous teams that have received AASHTO and NPHQ awards. In 2011, Webb was a recipient of the coveted MDOT Director's Award. She regularly sought to provide career perspectives for women in engineering, in general, and particularly people of color in related fields. In March 2016, as a featured guest on "American Black Journal," which was broadcast on Detroit Public Television, she spoke of her 25-year career and pioneer journey at MDOT and also highlighted her involvement in COMTO. Webb's contributions to the transportation engineering profession were cemented when, in March 2016, she was recognized during COMTO's 2016 Women Who Move the Nation Breakfast Awards as the Chairman's Eagle Award Honoree in Washington, DC. Webb holds a B.S. in Civil Engineering from Valparaiso University and is a registered professional engineer in the State of Michigan.

MOVE: Mobility Optimization through Vision and Excellence

Greer Johnson Gillis, PE, Vice President–System Development, Jacksonville Transportation Authority
Jacksonville, Florida

Background

Founded in 1955, the Jacksonville Transportation Authority (JTA) was established as an independent state agency serving multimodal needs for Duval County, Florida. Responsible for bridge and highway design and construction, bus service, the Skyway monorail, paratransit, and on-demand transportation services, the JTA serves the largest city in the continental United States with an integrated transportation network. In 2018, the JTA Board of Directors approved the Mobility Optimization through Vision and Excellence initiative, or MOVE. MOVE was intended to help JTA integrate new transit technologies and mobility alternatives into its business model. JTA also sought to become a regional mobility leader by supporting a robust regional transportation system that relies on a combination of traditional mobility solutions, new innovations, technologies, and partnerships. If only leaders had known at that time how important their forward-thinking would be to position the JTA for what was to come two years later.

In 2020, the nation was experiencing unexpected challenges as the global pandemic significantly reduced agencies' ridership by as much as 95%.[1] In tandem, social unrest and protests have swept through cities in a way not seen since the 1960s. As unparalleled numbers of Americans are home for work, for family care, or due to health concerns, the socioeconomic disparities grow more acute. With the vision to provide "Universal access to dynamic transportation solutions," JTA found itself at a crossroads. While the initiative and resilience of the JTA leadership and staff have carried the JTA and its employees, customers, and clients through these historic disruptions with remarkable success, it is necessary to pause and plan for the longer term and a new normal. These circumstances led to a strategic planning process and phased approach for MOVE.

Goals and Objectives

The phased approach and strategic plan development process provided opportunities for JTA leadership to navigate what it meant to provide transportation and transit services when America's attention was focused on the pandemic, the economic crisis,

---

[1] https://www.washingtonpost.com/transportation/2020/04/23/with-ridership-down-95-percent-losses-700-million-amtrak-looks-pandemics-recovery-phase/

and civil unrest. The goal of the refined strategy was to develop a resilient and sustainable roadmap for JTA to chart a path forward to the post-COVID future.

The MOVE Plan Phase I was based on JTA's unique strengths paired with industry trends to provide recommendations that focused on critical elements that include sustainability, resiliency, operational turnarounds, equity, and innovation for all areas of the JTA and their services. The Plan focused on immediate, tactical actions and provided a set of scenarios to define where the JTA may need to pivot in response to circumstances related to COVID-19 recovery, economic recovery, and local conditions over the next 18 months. MOVE Phase I prioritizes core customers (the transit-dependent and essential workforce who rely on transit), and the JTA also needed to creatively work through partnerships to advance equity, environmental sustainability, and technology in Northeast Florida.

The Plan revolves around three key themes:

1. Know Your Core: Address core customer needs equitably with reimagined services.
2. Build on Your Strengths: Lead the region in mobility management.
3. Collaborate for Success: Become a strong regional capital development partner and engage new partners for transit innovation.

The Plan also provided recommended actions to support organizational resiliency and continued growth and enhancement to the overall effectiveness and innovative spirit of the organization.

The Phase I Plan will serve as the foundation for further exploration in the development of the Phase II MOVE Plan, which will be the JTA's next five-year strategic plan (Figure 4.2).

FIGURE 4.2 MOVE Phase I brand and imagery used for the project.

Approach and Challenges

Through an assessment of urban mobility industry trends, the MOVE Plan considers national trends through the lens of Northeast Florida, forged by a partnership with the City of Jacksonville, and through intentional outreach and engagement to both staff and customers. The Plan combined available data, research, and survey responses from JTA staff and customers, advice from industry thought leaders, and insights from business trends. Exploratory scenario planning was used to address three time frames of recovery (COVID-19 recovery, economic recovery, and local conditions). By analyzing triggers (specific trend indicators that signal when events are aligning with a particular scenario) and proposed responses, JTA was able to develop a roadmap that would allow them to pivot or accelerate as events occur, thus strengthening the JTA's effectiveness as future events unfold.

Results

A tactical 18-month "Roadmap" was the result of this phasing and strategic exercise. This roadmap will identify new strategic directions and actions for JTA to take: conducting a thorough Route Reimagination Study to address changing transit needs and employment and service access. It also aided in developing and marketing new metrics that will allow the JTA to move away from revenue-based measurements and support the concept of operating transit as a public good, supporting the community through a vaccine distribution initiative using various transit services, and developing pilots and tactical urbanism to test opportunities for pop-up improvements in low-income, transit-dependent neighborhoods. The Plan also recommends actions that support growth and effectiveness of the JTA while continually working toward organization resiliency.

The MOVE Plan is the beginning of the JTA's strategic planning activities and was envisioned by CEO Nathaniel P. Ford Sr. He tasked his entire executive leadership team to be engaged in every facet of the development of this action plan. The project was led by VP Greer Johnson Gillis and VP Bernard Schmidt of the Automation and Innovation Division. The project executives oversaw a 20-person project leadership team consisting of senior managers across the JTA to provide guidance and insights in the development of the plan (Figure 4.3).

FIGURE 4.3 Two women near a shuttle used in the program.

About the Team

Greer Johnson Gillis is VP of System Development for the JTA. During her 27-year career, Gillis has implemented sustainable policies, managed large-scale infrastructure projects, and advocated for diversity within the engineering, energy, and construction industries.

Gillis joined the JTA in April 2020 overseeing Construction and Engineering, Planning and System Development, and Facilities Management. She also focuses on increasing workforce development initiatives and small business participation in the city. In her first months at the JTA, Gillis established the JTA's Capital Project Office to focus on efficient project delivery and initiated the first phase of the JTA's strategic planning efforts.

In 2014, Gillis was honored by President Barack Obama and U.S. Transportation Secretary Anthony Foxx as a White House Champion of Change: Transportation Ladders of Opportunity for her efforts in expanding diversity in the engineering industry. In 2017, she was honored by COMTO with the Women Who Move the Nation Local Government Award. In 2008, she was appointed by the Interim Director of District (District of Columbia [DC]) DOT to serve as the city's transportation lead for the 2009 Presidential Inauguration.

Prior to serving at the JTA, Gillis held several executive positions in the DC Government, including Commissioner on the Public Service Commission, Director of the DC Department of General Services, and Deputy Director of the District DOT. She also has many years of private sector experience working for several engineering companies.

Gillis is an active member of the Women's Transportation Seminar (WTS) NE Florida Chapter, and the COMTO Jacksonville Chapter. She holds a B.S. in Civil Engineering and a Master of Science (M.S) in Civil Engineering with a concentration in transportation, both from the Georgia Institute of Technology. She is a licensed professional engineer in Washington, DC, and Virginia.

Open Access Electric Vehicle Charging

Bonnie Datta, Former Senior Director, Regulatory Affairs and Market Development, Americas and Southeast (SE) Asia/Siemens; Current Managing Partner, Plug to Grid Strategies
California

Background

The idea of owning an EV is exciting to some, but daunting to many. Alternative-fuel-driven vehicles are being adopted at a slower pace in the United States in comparison to the European Union (EU) or China, with commentators attributing the adoption curve to both the lack of a unified national policy enforcing greenhouse gas (GHG) emission reductions and consumer behavior. The biggest concern expressed by consumers is the lack of EV charging infrastructure.

This worry (which often leads to the decision not to purchase an EV) comes from a combination of range anxiety and charger anxiety. Range anxiety refers to the availability of charging stations where needed along a route. Charger anxiety refers to the worry about whether the driver can use the station once reached. Adequate charging infrastructure is needed to provide driver confidence and combat range anxiety, but despite numerous government agencies providing funding to build out infrastructure, efforts in California were previously uncoordinated and not part of a long-term, strategic plan.

Charger anxiety may result when chargers are available along the routes for EVs to recharge, but the driver cannot access them because they are not a member of the charging network. In an attempt to protect proprietary business models, charging service providers deliberately try to lock in their customers by requiring them to join the provider's membership network and use proprietary radio-frequency identification (RFID) cards to access the chargers. This is not a viable system for customers, as poor cell service or a low-on-charge smartphone may prevent payment in such situations, making it impossible for the driver to use the charging station.

Goals and Objectives

As a company that strongly supports decarbonization with a commitment to net zero carbon by 2030, and as a recognized leading technology company, Siemens was well positioned to tackle this challenge. Additionally, Siemens is seen as an industry leader for their longstanding commitment to open standards and interoperability by promoting open payment standards.

Open Access Electric Vehicle Charging was a focused initiative addressing both range and charger anxieties in California. For charger deployment, the goal was to establish a process to create a strategic, statewide plan for charger deployment. Most of the funding for deploying chargers came from or was managed by state agencies such as the California Public Utility Commission (CPUC), California Energy Commission (CEC), and California Air Resources Board (CARB). Therefore, legislation was needed to establish a coordinated planning process across the state agencies.

For open payment access at public charging stations, legislation was already in place but was ineffective, allowing proprietary systems to dominate. CARB had instituted a rulemaking to make the existing legislation effective, but opponents sponsored new legislation to overturn CARB's process. Therefore, promoting open payment access required both supporting CARB's rulemaking and opposing the relevant proposed legislation.

Approach and Challenges

To resolve the lack of interagency coordination in charger deployment, an industry alliance, led by Bonnie Datta, worked to find a legislative sponsor and drafted legislation (AB2127). Unlike many EV bills that relied on support from primarily environmental groups, Datta determined that it was important that the legislation was seen to be widely supported by the industry and other stakeholders. She worked to achieve widespread support by building a coalition across EV supporters, including other industry alliances, environmental groups, electric utilities, consumer groups, and

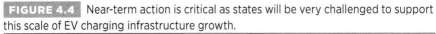

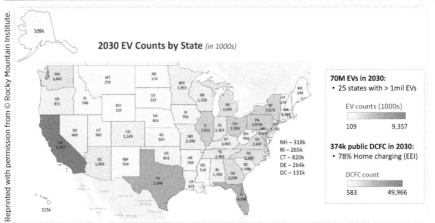

FIGURE 4.4 Near-term action is critical as states will be very challenged to support this scale of EV charging infrastructure growth.

others. The key challenge for her to overcome was demonstrating a need for the bill, but Datta and the coalition provided such a high level of education to key legislators and other stakeholders that this never became an issue.

In contrast, the initiative around open payment access faced huge challenges. Significant obstacles were raised by charging service providers working to protect their proprietary business models. California legislation in 2013 (SB454) required open payment access, but even by 2018, the industry's payment systems remained closed. CARB initiated a rulemaking to determine whether public charging stations should accept credit, debit, and prepaid cards without signing up for the network provider's service. Datta participated actively in the rulemaking and provided extensive technical and cost data to CARB staff in support of the open payment option. Datta utilized a coalition strategy in the proceeding with CARB, enlisting EV alliances and environmental groups. This approach was critical to the success of this initiative, as virtually all the other charger manufacturers and service providers in the proceeding fought against open access—often providing inaccurate or misleading data.

As CARB was nearing adoption of the regulation, opponents found a legislative sponsor to introduce a bill misleadingly entitled, "Electric Vehicle Charging Stations Open Access Act" (AB1424). In fact, the bill would have specifically prohibited CARB from adopting the exact regulation that it had decided was necessary to protect EV drivers (Figure 4.4). The challenge here was enormous because the bill's supporters moved quickly and passed the State Assembly before open access supporters had the ability to engage. This was one of many bills that coalition members were following, and these were not always high on their radar, but Datta kept this work high profile, working with the coalition, convening update sessions, driving efforts to submit opposition letters, and educating partners, legislators, and other stakeholders.

Results

AB2127, the bill to coordinate agency planning for chargers, passed with overwhelming support and is now being implemented by the CEC. AB2127 requires the CEC to biennially assess the electric vehicle charging infrastructure needed to meet the state's goals of putting at least 5 million zero-emission vehicles (ZEVs) on California roads by 2030. ZEVs are vehicles that produce zero exhaust emissions of any criteria pollutant (or precursor pollutant) under any and all possible operational modes and conditions. California accounts for approximately 50% of total national ZEV sales (CEC Feb 2021).[2]

Types of ZEVs include:

- BEVs: Battery Electric Vehicles, which run exclusively on batteries.
- PHEVs: Plug-in Hybrid Electric Vehicles, which have a limited electric range and a range-extending gasoline engine.
- FCEVs: Hydrogen Fuel-Cell Electric Vehicles, which are powered by electricity stored in hydrogen fuel.

Regarding open payment access, the combination of providing credible supportive information and Datta's leadership of the coalition gave CARB the support and evidence it needed to adopt the open access regulation.

Regarding the bill to overturn CARB's open payment regulation, Datta championed the effort to bring a wider coalition together and keep it focused while undertaking the necessary steps to ensure that the bill would get a fair hearing—which it did and, as a result, was defeated.

The Open Access initiative achieved both of its intended goals. While range anxiety will continue to be an issue, the CEC is now developing strategic plans for statewide charger deployment and ensuring that funding across agencies—including utility funds authorized by the CPUC—are being spent as effectively and efficiently as possible to create the needed charging infrastructure.

Regarding open payment access, CARB regulations are now in place requiring that all DCFCs installed at public locations be equipped with credit card readers starting January 1, 2022. The start date for Level 2 chargers, common at workplaces and retail sites, is July 1, 2023.

Other states are now considering similar policies.

About the Team

Bonnie Datta is passionate about emerging clean technologies and intelligent infrastructure. Datta is an international market strategist who brings to bear a combined policy and business background to positively impact decarbonization pathways, with deep expertise in transport electrification and grid modernization.

Armed with a multidisciplinary, cross-industry skillset, Datta is a policy expert with advocacy depth engaging policymakers and industry stakeholders globally. She is the founder of Plug to Grid Strategies, a niche advisory firm that works with a very select group of high-potential climate tech startups, with a focus on the autonomous, electrified, and connected transport ecosystem, and distributed energy resources grid connectivity.

---

[2] https://www.energy.ca.gov/data-reports/energy-insights/zero-emission-vehicle-and-charger-statistics

Prior to founding her firm, Datta was Senior Director for Regulatory Affairs and Market Development, Americas and SE Asia, Siemens Digital Grid and eMobility, where she was the Siemens "voice" engaging with policymakers in the advancement of smart energy policy positions and leading key customer accounts. Before she moved to the United States, she worked at the elite Singapore government agency responsible for the nation-state's inward investment programs—a pivot from her earlier position in mobile telecom managing product marketing for 15 countries in the Asia Pacific region.

Datta holds an M.A. from Jawaharlal Nehru University and B.S. from Loreto College. She serves on the Board of Interstate Renewable Energy Council.

5

Marketing and Communications

Steve Lagreca/Shutterstock.com.

Women are especially qualified in marketing and communications to discover, define and articulate the benefits of a product, place, or method of transportation. Rather than focusing on features, women in the field look at how to impact behavior in a way that changes outcomes for the greater good.

—Christina Lovio-George, CEO and Founder LovioGeorge Inc.
Communications + Design

As two professionals who have made their careers in marketing and communications, we have often found ourselves surrounded by other women—from our college classes to the office. Contrary to other sections of this book, women are the overwhelming majority of marketers today.

According to the Alliance for Inclusive and Multicultural Marketing, the workforce skew is 63% female versus 37% male among U.S. marketing departments.[1] With that said, however, the same study finds that women still lag behind men in achieving leadership roles. In fact, of the surveyed marketers, 45% have reached a Chief Marketing Officer (CMO) or CMO-equivalent roles, and only 46% have reached what they considered "senior level" roles in their departments.

---

[1] https://www.ana.net/blogs/show/id/mm-blog-2018-11-ana-advertising-diversity-report

Our careers have operated within the marketing and communications realm for quite some time, and despite the uphill battle, we have both been able to experience some extraordinary aspects of this industry. Although we carved different paths into roles in automotive and mobility, we both ultimately fell in love with one aspect: auto shows.

Being based in Detroit, we naturally gravitated to the North American International Auto Show (NAIAS), which is like the Super Bowl of automotive announcements and glitz and glam. The show was so much more than just about cars: it was a symbol of the future, a trophy case of the best ideas and technologies the industry had to offer, and, most important, it was a place where a journalist, an engineer, a philanthropist, a child, and a car enthusiast all wanted to be. The NAIAS and all of its moving parts craftily and strategically layered brand messaging, lifestyle curation, and market segments into a transformative interactive showroom experience. Regardless of industry, it would be any marketer's dream.

It is no secret that the industry is changing and shifting to different modes of transportation. We are seeing new materials being used in vehicle manufacturing and different propulsion systems powering vehicles. And, of course, there is a big focus on next-generation technologies and options like autonomous and connected vehicles and micromobility modes that are all working toward more sustainable, equitable, and safer transportation for all.

But these new innovations will only be successful if consumers are on board—a responsibility that falls primarily on marketers to drive sales.

The science of marketing and communications is understanding every audience, anticipating their behaviors and spending trends, and involving them in the conversations. Marketing and communications professionals will have to channel their special skillsets, which allow them to empathize with consumers; educate and show value; and anticipate consumers' wants, needs, and questions to address them specifically and directly. These skills can be used to drive vehicle sales, the adoption of public transit, compliance with safety regulations, and even to bring awareness to alternative transit modes.

As marketers and communicators, we are constantly thinking about the best message, the most effective graphic, and how to stay concise, personify places and things, breathe soul into products, show impact, and ultimately connect with the audience in a way that is authentic to everyone. The emotional relationship between brand and person makes all the difference.

The following stories will showcase the transformative work that can come from the right words, the right story, and the right visuals. Each case study showcases the magic of the automakers, beyond what they have done at the NAIAS. Each of these stories showcases the power of the female perspective and the importance of diversity and inclusion, especially when hoping to attract a new customer base.

Bringing Back Bronco, 2021

Jovina Young, Bronco Sport Brand Manager, Ford Motor Company
Lindsey Laporte, Bronco Experience Manager, Ford Motor Company
Erica Martin, SUV Communications Manager, Ford Motor Company

Darci Gurney, Brand Content and Alliances Manager, Ford Motor Company
Lisa Schoder, Media Manager, Ford Motor Company
Courtney Nowicki, Media Manager, Ford Motor Company

Background

The Bronco was originally engineered in 1964 by Paul G. Axelrad and with the approval of Lee Iacocca—an influential and iconic Ford Motor Company executive—to serve as an off-road vehicle. It went through five generations as the American sport utility vehicle (SUV). The Bronco would be discontinued in 1996 when it was replaced with the Expedition (Figure 5.1).

At the 2017 NAIAS, Ford confirmed that the Bronco would return for a 2021 model year release. The new model would be based on the Ford Ranger, but with first-generation body features that would further off-road capacity.

The relaunch of the Bronco is a key component of Ford Motor Company's trajectory on future vehicle models and as a company. Jeep is a longstanding competitor of the Bronco and has been the sole market leader in the off-roading space (Figure 5.2).

Jovina Young, Erica Martin, and Darci Gurney were key players in the relaunch of a product and brand that had been out of the market for more than 25 years. The stakes were high. This rebrand and relaunch was a huge opportunity for Bronco to reclaim their leadership in the space that Ford walked away from with the discontinuation of Bronco in 1996. Internally, the Bronco was a symbol of how Ford was effectively challenging the status quo. The high anticipation from customers added an additional layer of pressure as they have been anxiously awaiting the return of the Bronco. The team needed to ensure the Bronco brand stood true to what their customers knew and loved about Bronco while also bringing it to a new modern meaning (Figure 5.3).

FIGURE 5.1 2021 Bronco Family: (left to right) Bronco Sport, Bronco 2-door, Bronco 4-door.

FIGURE 5.2 Darci Gurney attends Bronco Off-Roadeo near Austin, Texas with Team Bronco ambassadors.

FIGURE 5.3 Jovina Young (left) and Erica Martin as they compete with a Bronco Sport in the 2020 Rebelle Rally.

Goals and Objectives

1. Establish Bronco as a capable and durable family of rugged SUVs: having been out of the marketplace for more than 20 years, they needed to quickly gain credibility in the segment (Figure 5.4).
2. Build an aspirational lifestyle brand to attract a broad range of customers: the team knew that Bronco would naturally appeal to Bronco and off-road enthusiasts, but they also knew that they wanted to be inclusive and lean into lifestyle elements to give the Bronco brand longevity and make it a cultural icon.
3. Support and build momentum for the Ford master brand: Bronco is a key proof point of Ford's trajectory and needs to strengthen Ford Motor Company's appeal for those who have not owned or considered Ford in the past.

Approach and Challenges

First, the team settled on our "Built Wild" positioning, which is very fitting given that a bronco is a wild horse by definition. This positioning also allows the brand to easily lean into the outdoor lifestyle space, encouraging people to get out and enjoy. They began teasing this positioning through their new Instagram handle, @fordbronco.

Next, after keeping Bronco under wraps for two-plus years, they needed to lift the sheet and share the Bronco family with the world. The plan was set: reveal in June

FIGURE 5.4 2021 Bronco.

2020 at the Detroit Auto Show—but when a global pandemic hit, it was clear they would need to change course.

The Bronco team knew they needed to embrace the cultural challenges and not shy away from it. People were feeling closed in because of stay-at-home orders and social distancing, and they were craving connection to the outdoors and the restorative elements of nature. Ford wanted to help give access to the wild in a way only Bronco could, at a time when people needed it most—even if it was while they sat at home.

So, in these unconventional times, they took an unconventional approach by connecting this iconic vehicle with an iconic media partner. A partnership with Disney provided mass reach, and Disney's broad range of networks and access to talent allowed Bronco to stay true to their Wild positioning. In partnership with Disney CreativeWorks, Ford created three short films that fit contextually within Disney programming and aired them simultaneously on one night across multiple Disney networks. Treating these films like full-length movies, the Ford team teased them for one week, creating momentum with social content, and then followed everything up with the first live look during "Good Morning America" the following day. For enthusiasts, there was a 9-minute reveal video including all of the product detail they could ever want. This multichannel approach generated interest and reach that even an auto show event could not rival.

On the heels of a wildly successful reveal, the Bronco brand continues to work toward their objectives through continuous human-centered product innovation and lifestyle extensions:

- Regular input from key enthusiasts and our Bronco Expert panel, comprised of non-Ford off-roading specialists.
- Ongoing torture testing and participation in desert racing events, including Baja 1000, King of the Hammers, and Rebelle Rally.
- The Bronco Wild Fund, dedicated to reforestation, trail management, access to the outdoors, and scholarships through nonprofit organizations, like Outward Bound, the National Forest Foundation, and Sons of Smokey.
- The Bronco Off-Roadeo, a unique consumer experience and destination for Bronco owners to learn off-roading skills, connect with experts, connect with one another, and enjoy the outdoor adventure lifestyle.
- Bronco Amazon Storefront, a place for Bronco fans to find everything they need in order to represent.
- Partnerships with like-minded brands and ambassadors, like Filson, Marmot, and Moment Lens.

Results

The Bronco family reveal trended on Twitter, YouTube, and Facebook during reveal week, and reservations (preorder) goals were wildly surpassed. This success came as a result of keeping a customer-first mindset and using flexibility to connect with them.

As off-roading segments tend to be masculine and male-centric, having strong female representation on the team was a crucial part of creating an inclusive brand and reaching more than this initial male audience. The introduction of Bronco Sport—a more compact SUV under the Bronco nameplate—provides a big opportunity to connect with women in the small SUV space, which is the largest segment for women. Thanks to the strong foundation set by a successful reintroduction, the Bronco brand will live for many decades to come, continuing to be a leader of innovation in the space while keeping true to the original brand missions and their roots.

About the Team

Jovina Young, Bronco Sport Brand Manager, was responsible for the overall marketing strategy and planning for Bronco Sport, including product plans, promotional alignment, pricing, and distribution strategy.

Lindsey Laporte, Bronco Experience Manager, was responsible for the strategy and execution of the Bronco Off-Roadeo consumer experience and communications marketing.

Erica Martin, SUV Communications Manager, was responsible for the Bronco brand family communications strategy and content creation.

Darci Gurney, Brand Content and Alliances Manager, was responsible for the strategy and execution of alliances, brand content, and nontraditional media, including the creation of Bronco Wild Fund, experiential, events, media integrations, outdoor enthusiast/brand partnerships, and more.

Lisa Schoder and **Courtney Nowicki**, Media Managers, are both are responsible for the media plan to drive awareness among key consumers—led key efforts to partner with Disney in the reveal plans.

"No Way, Norway" Superbowl LV Campaign, 2021

Courtney Smith, Digital Strategist, General Motors

Background

In January 2021, GM debuted its new brand identity and marketing campaign "Everybody In," signaling its rallying cry for widescale adoption of EVs. This messaging was positioning GM's Ultium Platform to be a key tool to lead the industry to widespread EV adoption. The goal of the Everybody In marketing campaign was to provide a call to action to the global community that they can, together, work toward a zero-emission world while highlighting GM as a leader in this moment.

With 30 new EVs launching globally by 2025, GM's communications and marketing teams were tasked to help grow awareness and excitement for EVs. To do so, they leveraged one of television's biggest audiences: the 2021 Super Bowl.

Goals and Objectives

EVs and the benefits they will bring to society are serious topics, but the overarching goal of GM's campaign was to use humor to curate a global conversation about EVs. GM's Super Bowl ad featured actor Will Ferrell, who discovers that Norway far outpaces the United States in EV adoption. In the commercial, "No Way, Norway," Kenan Thompson and Awkwafina join Ferrell on his action-packed journey to give Norwegians a piece of his mind. Reflective of the overall campaign goal, the Super Bowl ad strategy was developed with the hope of sparking a conversation and a movement, inspiring others to join in.

Aside from using humor to launch a global conversation, GM sought to:

- Create awareness of the Everybody In campaign and the importance of EVs in creating a cleaner, safer world.
- Secure a top spot in Super Bowl ad roundup coverage and beat internal earned media benchmarks for past Super Bowl activations.
- Deliver an exciting content ecosystem that positions GM as a fresh, modern, and approachable tech company.

Approach and Challenges

Today, well over half of new vehicles sold in Norway are electric. Comparatively, EVs made up fewer than 4% of the market share in the United States in 2020. Knowing this, GM used the idea that Americans do not like losing to fuel the campaign. The communications team worked closely with Weber Shandwick and McCann Worldgroup with marketing support from FleishmanHillard, GM's agency of record. Once Ferrell was secured for the role, he led a lot of the script development and brought on his director to make sure it was authentic and that the tone was delivered in a positive, fun way.

Multiyear analytics show that brands that fully reveal their ads before game day achieve greater earned attention, social interaction, organic digital engagement, and overall stronger impacts. Most media coverage of Super Bowl ads is published from Monday through Thursday of Super Bowl Media Week, and GM's approach was to reveal early and build buzz throughout the week.

Using an owned, earned, and paid digital strategy (ensured that the Super Bowl ad was not just 1 minute during the game, but instead a weeklong buildup to the game that fostered the growth and reach of the Everybody In campaign messaging, laying a foundation for the campaign's ongoing efforts.

Paid digital and social advertising also launched across Facebook, Instagram, and Twitter aimed to build excitement with teaser video ads. Ads were strategically targeted at GM's core audience of policymakers, investors, potential employees, and "modern citizens"—a custom-made audience segment reflecting people who are passionate about transportation, the environment, and their communities. Organic social posts on GM's channels brought the campaign messages, engagement opportunities, and deeper education to the brand's existing followers.

The initial messaging created a bit of a challenge to the team as they addressed audience sensitivity. Early development messaging for the campaign included lines like, "We're coming for you, we're going to punch you in the face." It is shocking for GM to take this stance and lead with this sort of messaging, and the team was hesitant and nervous that it could be misinterpreted.

Although the company knew this approach was bold, they wanted to be crystal clear that the messaging was not inciting violence, nor threatening anyone. Through this internal discussion, the team connected with the GM international relations team to give them a courtesy preview of the commercial. This led to talks with Norwegian Prime Minister Erna Solberg; the Ambassador of Norway, Anniken Krutnes; and Norwegian EV advocacy groups. While this proactive approach was expected to be seen as an olive branch to the team, it went better than expected. The relationship between GM and Norway turned into an ongoing conversation through which Norway was able to use this opportunity to showcase their success in EV adoption. This collaboration will continue to develop educational opportunities for the global conversation on EV adoption.

Results

This campaign beat both internal and external expectations, from an engagement standpoint. Traditional Super Bowl advertisers hold commercials until game day. In doing so, the opportunity to earn engagements is reduced. With the knowledge that most commercials are not really discussed following the game as much as they are in the days leading up to it, GM took the opportunity to release their ad exclusively to NBC to air the Wednesday before the Super Bowl. In tandem with this, GM pushed out previews to other ad publishers and automotive/tech journalists. A few select interviews were given for a behind-the-scenes look with the *Detroit Free Press*, *Adweek*, and *Ad Age*.

GM's tweet revealing the ad became GM's highest-reaching, most-engaging organic tweet ever. Over the next few days leading into the Super Bowl, GM's social mentions averaged 3,200 a day, increasing 10 times on Super Bowl Sunday. GM received reactions from prominent groups; influencers; leaders such as the Ambassador of Norway, Judd Apatow, Congressman Tim Ryan, Secretary General of the Norwegian EV Association; Audi; Uber; and even Norwegian universities.

Earned media metrics:

- GM's Super Bowl activities generated 2,431 online and print news stories with an Estimated Earned Reach of 23.2 million.

- There were 1,725 broadcast clips collected with a combined Potential Viewership and Station Reach of 97.6 million.

- GM's "No Way, Norway" ad was listed as a "Top Super Bowl Ad" by *Tom's Guide, Hola!, Adweek, Business Insider, Washington Post*, AP, *TooFab, USA Today, Bleacher Report, Collider,* CNN, *Digg, HuffPost, Forbes, Vogue, ET, Billboard, BroBible, People, SportsNet, The Athletic, Looper*, and others. The "No Way, Norway" ad also appeared in the *Axios AM* newsletter as the "Best Commercial" during the Super Bowl.

Owned media results:

- To date across all organic social posts, GM has generated 68,700 engagements, 282,800 clicks, and 9.09 million reach, with 2.52 million video views.
- GM's Super Bowl presence lifted GM's electrification visibility daily average 225%.
- There was a significant conversation volume spike on Twitter after the ad ran, with mentions up 253%.
- Two-thirds of GM's Super Bowl engagement and more than three-quarters reach was driven by the full ad reveal alone, namely, from Twitter, which was also the channel's best-performing organic tweet ever.
- The teaser video series opened a unique opportunity for the GM brand to engage with various policymakers, government leaders, Norwegian EV associations, and tourism organizations—U.S. and Norwegian—to build on the true #EVerybodyIn momentum on social media. We welcomed the healthy competition from other automakers who created original content in response to our ad.
- On game day, GM used its social presence to spark an EV dialogue and to engage with social audiences including celebrities and influencers.
- During the Super Bowl, "No Way, Norway" was trending in the United States on Twitter.

Additionally, Will Ferrell is expected to have a GM EV by the end of 2021.

About the Team

Courtney Smith leads digital strategy for the GM brand communications team. In this role, she is responsible for the editorial strategy on GM.com, as well as supporting multiple content partnerships across the digital and social influencer space. Prior to this role, she led the corporate social strategy through several crises and started her career at GM in 2015, working in manufacturing communications and supporting several plant communities. Smith has a bachelor's degree in public relations from Wayne State University and is pursuing her master's degree in data marketing communications at West Virginia University.

Multicultural Marketing "Built Phenomenally," 2020

Rajoielle Register, Head of Global Brand Experiences, Ford Motor Company
Laura Busino, Head of Content Production, Ford Motor Company
Dibrie Guerrero, Multicultural Marketing Communications Manager, Ford Motor Company

Background

Ford Motor Company, like all automakers, must continue growing its customer base to remain a top competitor in the automotive market. For the Ford Escape, one of the most important customer segments was identified as African American women. Research has shown that African American women are often the decision-makers whether they are head of household or not. Their influence in their homes and communities is often higher than that of the general market. Knowing it was imperative to reach this audience target, a strategy was developed to create a 360-degree view of "her" the African American female audience.

Ford prides itself on being a company based on human connection and purpose. Our core values lead us to strive for authentic human connections with our customers. With more than 50% of Escape registrations being by African American women, we needed to create a body of work that truly represented her and reflects what this vehicle represents for her (Figure 5.5).

Goals and Objectives

It was important to have the all-new 2020 Ford Escape stand out from an overcrowded segment of compact SUVs whose storylines are similar. The Ford team sought to be impactful, to have relevance, and to build awareness and consideration through brand connection with the audience. Additionally, the team realized they had a unique opportunity to intentionally tell her story by demonstrating how Ford understands and connects with this customer base. Understanding that representation matters, particularly with African American women, Ford wanted to make it matter both in front of and behind the camera.

As the first founding automotive partner of Free the Work (a talent-discovery platform for underrepresented creators) and one of the first to implement the Association of National Advertisers' (ANA) SeeHer Gender Equity Measurement ad scoring, Ford had the tools and structure in place to make an organizational impact. Both commitments allowed the automaker to be comfortable using that mentality to push female representation—specifically black female representation—across all contributors to this project.

Approach and Challenges

Building authenticity in an ad campaign seems almost inherently oxymoronic. To make sure Ford avoided that pitfall, the team started from the ground up. They did not just have an African American woman write the script and hand it off to a white, male director to shoot. For most of those involved in the process, from the creative idea through the execution, this was an authentic story. It was created out of their own experiences as Phenomenal Women. The campaign was called "Built Phenomenally," based on the poem by Maya Angelou. The creative concept itself was an inside/outside look at how a commercial is created. In this meta-fashion of a commercial about the filming of a commercial, Ford was able to highlight the production choices and the representation and equity programs that the company was already championing.

FIGURE 5.5 "Built Phenomenally" team.

Ford also decided to include teenage girls in the production process to show them the art of the possible. Following the ANA SeeHer motto, "If you can see her, you can be her," Ford partnered with Made in Her Image and Girls Make Beats to create a platform to mentor our next-generation girls.

Results

Built Phenomenally was deemed an overall success and was considered transformational among the intended target. Brand Favorability is +4.2 point Lift, 70% higher than U.S. Auto Vertical Norm. (+2.5 point Auto Vertical Norm and +1.7 point NA Norm) on Facebook. The ad also resonated with and gained coverage on several top African American media outlets.

While the Escape project is complete, the need to tell authentic stories and connect with a wide customer base is not. What this project proves is that how you create your content affects the content itself. This is not to say that an empathetic person cannot tell another person's story, but think about your own life: Think about some of the very small, specific details about being a parent, a woman, or a teenager, or dealing with an ill family member, or living life on Zoom. Someone who had not had a similar or shared experience might not be able to add to your story.

Luckily, Ford has been building and will continue to build the infrastructure within their production process to grow and hire creative talent from diverse backgrounds. This is the only way we can tell a wide range of stories and be exposed to a wide range of experiences. Because women are the decision-makers for the majority of vehicle purchases, we need to make sure they are represented in an honest way.

About the Team

Built Phenomenally was a labor of love from a phenomenal team of women, both inside and outside of Ford.

As the creative agency, Ford brought on UniWorld Group with N'Jeri Nicholson, Writer; Sharon Kimbrough, Producer; and Kim Gormley, Digital Producer at the helm.

This project was also made possible with Kanyessa McMahon, Director at Mirror Films; Daisy Zhoi, Director of Photography; Lizzy Graham and Meg Kubicka, Editors at Whitehouse Post; Marci Rodgers, Stylist; Lola Okanlawon, Makeup Artist; Carrie Bernans, Stunt Driver; Angela Bassett, Voice Over Talent; and Big Freedia, Soundtrack. Additionally, Ford drew from the expertise of Girls Make Beats—a nonprofit organization that empowers girls by expanding the female presence of music producers, DJs, and audio engineers—and Made in Her Image—a nonprofit movement striving toward social equity in the film, media, and entertainment industry for girls and women of color.

The Ford Motor Company team included Lisa Schoder, U.S. Media Intelligence and Strategic Planning; Raj Register, Head of Brand Strategy and Growth Audience Marketing; Dibrie Guerrero, African American Marketing Communications Manager; Laura Busino, Head of Content Production; and Sara Hendricks, Content Production Analyst.

Rajoielle "Raj" Register is Head of Global Brand Experiences at Ford Motor Company and is a diverse and detail-driven ambidextrous thinker with a background in product development and marketing. She has a passion for understanding and developing consumer experiences that translate to moments that matter. In her current role, Register leads by collaborating with key stakeholders to deliver the corporate brand experience and marketing strategy for Global and Regional Auto Shows. Her role also includes implementing brand integration and aligning strategic priorities for key moments at experiences like CES, SEMA, NADA, and AirVenture.

Prior to joining marketing sales and service, Register spent several years working in Ford's body engineering and product development segment. Register has a Master in Business Administration (MBA) from Duke University's Fuqua School of Business, an M.S. in Engineering Management from Wayne State University, and a B.S. in Mechanical Engineering from Michigan Technological University.

Laura Busino is the head of content production at Ford Motor Company, where she oversees all U.S. production across Ford's agencies and develops the guidelines and rules to make sure Ford is always represented consistently. Laura comes from an agency background and understands the process of building creativity and authenticity from the ground up. She believes the "how" is just as important in the "what" and tries to implement that across her team and the work they produce.

Dibrie Guerrero is the multicultural communications manager at Ford Motor Company specifically dedicated to African American audience communication. She leads the strategic development and implementation of go-to-market actions to deliver upon the brand's dedication to its consumers and products, building customer loyalty, brand recognition, and customer satisfaction. Dibrie also is the client lead with the oldest U.S. African American agency—UniWorld Group—collaborating on leading and developing marketing communications through data-driven planning for Tier 1 advertising creative through all media channels. Through proactive storytelling, programs, and partnerships, her role serves as the voice of customers for Ford communications delivering Ford's mission and vision.

6

Mobility on Demand

Mobility as a service provides the opportunity for more equitable mobility options for more people—if done with a balanced, intentional approach. While the private sector's strength is in bringing technological advances in shared mobility options to market more quickly and cost-effectively, the public sector must continue to play a role in ensuring those options are distributed equitably, with an eye on providing access for those most in need of a variety of mobility options.

—Lisa Nuszkowski, President, M-1 Rail and Executive Director, MoGo Bikeshare

Today, the norm of transportation is evolving, especially in the arena of individual car ownership, which has been waning to shared mobility options—a trend that has been gaining traction in recent years. This is especially the case in urban areas where congestion and inadequate parking bring frustration to car owners. Some individuals find owning a vehicle to be too costly or burdensome, while others wish to lessen their carbon footprint and seek more sustainable options as opposed to single-occupancy vehicle (SOV) trips. Additionally, some individuals are turning to privatized, shared-transit platforms as another alternative.

While car ownership might not become entirely obsolete, it will likely be less widespread and not considered a necessity among a large population of Americans. With this increase in users and change in focus by the masses, the mobility industry—along with transit agencies—is challenged with creating a system that fills the gap in transportation.

Mobility on Demand (MoD) refers to the ability of people to utilize varying transportation modes to make their journeys more efficient. This can be done by leveraging emerging technologies and facilitating public-private partnerships to allow for a user-centric approach. It is a relatively new concept based on the principle that transportation is a commodity in which modes of transit have economic values that are clear based on costs, journey time, wait times, number of connections, convenience, and other attributes.[1]

MoD is about providing travelers more seamless travel options through the integration of on-demand services and public transportation while simultaneously promoting public transit combined with first-mile and last-mile connections—meaning the first and last segments of a commute journey, such as the segment from home to the bus stop on your way to work. It encapsulates how people move, how households consume goods and services, and the spatial aspects of consumer decision-making. MoD represents an evolution of our transportation system that reflects changing sociodemographics and integrates innovative practices, solutions, and models for the management of transportation supply and demand.

The USDOT believes MoD should represent a vision for future mobility as a safe, reliable, and carefree mobility ecosystem that supports complete trips for all and is achieved based on several guiding principles[2]:

- User-Centric: Promotes choice in personal mobility and uses universal design principles to satisfy the needs of all users.

- Mode-Neutral: Supports connectivity and interoperability where all modes of transportation work together to achieve the complete trip vision and efficient delivery of goods and services.

- Technology-Enabled: Leverages emerging and innovative use of technologies to enable and incentivize smart decision-making by all users and operators in the mobility ecosystem.

- Partnership-Driven: Encourages partnerships, both public and private, to accelerate innovation and deployment of proven mobility solutions to benefit all.

---

[1] https://rosap.ntl.bts.gov/view/dot/34258
[2] https://www.its.dot.gov/factsheets/pdf/MobilityonDemand.pdf

Everyone needs the ability to travel more easily and safely.

This chapter identifies several case studies that showcase the full gamut of MoD, including public to private sector examples that reimagine the way people perceive public mobility. One story in particular hits close to home for us as it details how one of downtown Detroit's largest employers sought to change employees' perception of public transit and alternate modes of commuting, which has lessened the company's reliance on parking spaces, lots, and structures within the business district. With enough conversion in this field, areas formerly used to house empty cars for hours on end can possibly be repurposed as other community spaces—making for a more vibrant Detroit. Other case studies focus on the benefits of multimodal transportation and how choosing the right mode for specific commutes can change the landscape of cities.

Limited Access Connections

Penny Grellier, Community Development Administrator, Pierce Transit
Pierce County, Washington

Background

Despite encompassing the second-most populous county in the State of Washington—Pierce County and the Greater Tacoma region—certain areas within the Pierce County Public Transportation Benefit Area have limited or no fixed-route bus service. Some park-and-rides in Pierce County reach capacity early in the morning commute, and many residents live more than half a mile from a fixed-route bus stop. In many areas, infrastructure does not support safe, accessible walking or biking to reach transit, leaving people reliant on personal occupancy vehicles (POVs). Additionally, traffic on the local Interstate 5, near Joint Base Lewis-McChord, is the most congested in all of Washington State.

Pierce Transit covers 292 square miles with three service offerings: fixed route, shuttle paratransit, and vanpools, with a mission to reduce reliance upon POVs for residents by providing safe, reliable, innovative, and useful transit. Through the Limited Access Connections Project, Pierce Transit worked to develop innovative transit solutions and solve access challenges.

Goals and Objectives

Funded by the Federal Transit Administration (FTA) as part of its MOD Sandbox program, Limited Access Connections' goal was to improve the availability of shared-use mobility options between Transportation Networking Companies (TNC) and transit agencies.

A three-pronged approach was determined to provide riders access to transit in select zones:

1. Provide first-mile and last-mile connections to the nearest bus stop or transit center.
2. Provide rides home outside the span of service.
3. Provide rides to and from park-and-ride lots and local train stations to alleviate congestion.

To ensure no financial strain was placed on residents, the project aimed to provide transportation trips that were fully subsidized by grant funds.

Approach and Challenges

After receiving the MOD Sandbox funding in October 2016, Pierce Transit identified community partners who stood to benefit from this pilot with their riders. Partners provided ridership data, level of use, start and end points, reduction of parking lot usage, and service satisfaction. In 2019, Pierce Transit established partnerships with Lyft, Sound Transit, and Pierce College Puyallup to provide subsidized rides to specific populations who faced limited access to transit options.

Two major complications came into play when launching the project: collecting data and providing ADA-accessible options. After nearly a year of negotiations with the original partner, Pierce Transit was unable to reach an agreement on data sharing and, therefore, selected another TNC partner, delaying the project launch. Once in process with the new partner, the data received was not as originally planned due to privacy concerns. Additionally, the new TNC partner did not offer accessible service, requiring a search for another wheelchair-accessible provider. This was complicated, in large part, due to the unknown nature of the proposed service.

The project launched in May 2018 with initial trip requests gradually increasing over the first few months. The options provided by Limited Access Connections varied greatly from the traditional fixed-route service to which residents were accustomed. To increase community awareness and ridership, marketing and outreach became key parts of the project's success.

Results

The award-winning project was completed on December 31, 2019, and Pierce Transit is committed to furthering the success of transit in Pierce County, even in a post-pandemic environment. Pierce Transit announced that the agency is pivoting toward gradual service-level restoration and is committed to having transportation services anchor Washington's economic recovery as people return to activities outside of the home beyond 2020.

Over the entirety of the funded project, 313 unique users accounted for 8,827 trips, with an average trip duration of four miles per trip—resulting in an average trip cost of $11.47. In partnership with Lyft, the program fulfilled more than 8,000 rides to bus stops at no cost to residents.

Concurrently, Pierce Transit established a partnership with local public schools and community colleges, which provided free bus passes to all 7,900 high school students in the county, as well as an additional 8,000 college students.

During the life of Limited Access Connections, Pierce Transit had no requests for a Wheelchair-Accessible Vehicle (WAV) service by riders. This may have been because the rider would have to connect to fixed-route transit as part of the journey, and many WAV users had access to paratransit already or were using fixed-route service without needing first-mile or last-mile connection.

Of riders who presented anecdotal feedback, the majority found the Limited Access Connections service very useful, many citing personal mobility issues that made it otherwise difficult for them to access fixed-route transit.

Since the conclusion of the project, experiences and data gathered from riders have helped Pierce Transit design and implement a microtransit connector service called "Runner," which launched on August 1, 2020.

About the Team

In an effort to find first-mile and last-mile solutions, Pierce Transit's Planning and Community Development Department sought funding and partnerships and was awarded the MOD Sandbox grant. Experience with community engagement was critical to the success of this effort, which was led by Penny Grellier, Community Development Administrator for Pierce Transit.

Penny Grellier, though not a planner by trade, worked to solve the puzzle of a connected transit system for the Pierce County region. Grellier is now managing Pierce Transit's microtransit project.

Grellier works on ORCA (One Regional Card for All) business accounts and innovative projects for Pierce Transit. Her work with Pierce Transit, and previously with a local nonprofit agency, has made transportation connections easier for the public in the past two decades. She is an avid transit rider and serves as Pierce Transit's Employee Transportation Coordinator. In 2019, she received Downtown On the Go's Transportation Innovator Award for her work on the Limited Access Connections Project.

Bedrock's Alternative Mode of Transportation Campaign

Vittoria Valenti-Amodeo, Mobility Team Lead, Bedrock
Detroit, Michigan

Background

Detroit has seen an economic boom in recent years and has experienced rapid growth in the central downtown area. Within historic Motor City and its deeply ingrained car culture, employers downtown quickly learned that the increase in activity meant an increase in commuters entering the city and a greater demand for parking options.

Parking availability has become a dwindling resource in Detroit, perhaps because of the overwhelming commute culture of SOVs coming into the city on a daily basis. Developing parking garages is an option, but this is not the best use of land, it comes with a long lead time, and ultimately is not a sustainable solution.

One of the largest corporate growth drivers in the area has been Bedrock, a full-service commercial real estate firm specializing in the strategic development of urban cores. Bedrock's portfolio includes more than 100 properties totaling over 18 million square feet. As a part of the Rock Family of Companies, Bedrock's parking facilities serve nearly 20,000 team members as well as the general public.

As one of the city's largest employers, Bedrock and its family of companies wanted to find ways to be a part of Detroit's transition as a world-class mobility city as well as attract and retain top talent and companies to the city to share in the economic reinvestment and prosperity.

One of the ways to draw in talent is to provide a stable, reliable, and stress-free parking platform to alleviate the stress and constraints on parking options and availability. This requires shifting the sentiment away from SOVs and overall transportation behaviors in a city that culturally identifies with cars and driving.

To address the issues around parking availability, Bedrock set out to create a program that would change employee and visitor commute behavior and ultimately lessen the demand on overall parking supply.

Goals and Objectives

As Detroit continues to grow and develop, there are limitations to the number of available parking spots and structures. Building new parking infrastructure to meet this demand takes time, but altering commute and transportation behaviors can have a more immediate effect. With this in mind, Bedrock aimed to change commuter sentiment toward alternative commute practices by rewarding those who opted to use an alternative mode—such as biking, public transit, ride-sharing, and carpooling—thereby reducing the demand for individual parking spaces.

By implementing a commute program, Bedrock sought to shift the overall driving behavior of individuals and encourage them to at least try a new mode of commuting with the hope of creating long-term habit changes. Such changes provided flexibility for individuals to affordably, quickly, safely, and, in some instances, more enjoyably, get to and from downtown.

Approach and Challenges

At the time of the program's rollout in 2018, Bedrock was the first local real estate developer to try a commute program that incorporated a wide range of transportation options on such a large scale. To increase participation in alternative commute options, Bedrock provided its employees with a range of incentivized options that included subsidized passes to the city's light rail system (Q-Line) and the regional transit authority (DART).

Bedrock also worked to bring Scoop to Detroit for its first major expansion in the Midwest. Scoop, a carpooling startup, provided an easy-to-use platform to connect individuals traveling along the same route to work. This allowed travelers to connect and determine an ongoing carpooling schedule. In addition to serving Bedrock, Scoop was available for the larger Metro Detroit area, providing the general public with a new way of getting in and out of the city.

The largest challenge the program saw in changing behaviors revolved around educating team members on how to commute differently. Since car culture reigns supreme in Detroit, many team members had never utilized the bus system or were unfamiliar with bus routes that could efficiently get them from home to the office and back. A large internal educational and marketing effort was implemented to raise awareness among team members and to demonstrate an individual value proposition—what would eventually lead to individuals changing their commute habits.

Another challenge Bedrock encountered was anxiety over anticipating proper commute methods at the beginning of the day for the evening commute. Many expressed fear over missing evening buses due to last-minute meetings, the unpredictability of Michigan weather when relying on biking, or coordinating schedules with coworkers for carpooling. To alleviate such concerns, Bedrock worked with the ride-hailing service Lyft to offer complimentary rides for team members who ran into issues with or missed their planned alternative mode of transportation.

There was also the challenge of coordination and planning. To make commuting easier for team members, Bedrock partnered with Luum, a mobility infrastructure provider, to develop the MyCommute platform. MyCommute serves as a central resource to educate team members about existing and new commute options—essentially making all the options offered by Bedrock accessible on a single platform. This allows team members interested in carpooling or vanpooling to be paired with one another based on schedules and home locations. The platform even delivers seminars to educate and prepare team members who are interested in trying something new such as public transportation, carpooling, shuttles, and more.

Results

Within the first six months of the initial campaign, more than 3,800 team members had logged at least one commute using an alternative mode of transport. The major focus of the program is to encourage behavioral change, and since its launch, it has seen a substantial portion of the targeted audience increase individual usage of alternative modes of transport.

Impact is primarily measured by looking at bicycle, carpool, public transit, ride hail, vanpool, and walking commutes as a percentage of all trips logged and tracked in MyCommute. These alternative modes of transportation made up 14.3% of all commutes logged in the first 30 days of implementation. That percentage has steadily increased over the first year, rising to nearly 20% in June 2019. Bedrock created a culture in which individuals who have tried an alternative mode of transportation at least once are likely to continue to use such services on a regular basis.

This change represents an increase of 21,500 trips using alternative modes of transportation, resulting in an average reduction of 316 parking spaces used per day and helping to prevent more than 3.4 million pounds of carbon dioxide (CO_2) emissions by reducing the number of SOV drivers on the road.

Bedrock's program is highly regarded and was recognized by the World Economic Forum in 2019 in a report on the corporate mobility transport challenge. The success of this program reinforces the notion that if you remove barriers and develop a proper program and educational campaign, people will try something new and can establish new habits.

Additionally, this program provides an excellent base for Bedrock to continue its work in developing Detroit as a mobility city and developing a greener transportation system, while also allowing Detroit to redevelop parking areas to serve the community rather than house cars.

About the Team

Bedrock's alternative mode of transportation campaign incorporated diversity in thought, skillset, race, and gender to highlight new perspectives and ideas that Bedrock's mobility team had not encountered. It challenged others to rethink their commute habits and the technology that could enable Bedrock to better bring its goals to fruition.

Vittoria Valenti-Amodeo, Mobility Team Lead at Bedrock, is responsible for multiple parking and mobility initiatives at the organization. Formerly, Valenti-Amodeo managed the activation of Bedrock's public spaces, where she was the catalyst for creative space usage throughout Detroit focused on attracting and connecting people throughout the region and beyond. Additionally, Valenti-Amodeo has led Mobility Week Detroit (2018) and the Detroit Moves mobility festival (2017, 2018, and 2019).

FlexLA: FASTLinkDTLA, 2018

Hilary Norton, Executive Director at FASTLinkDTLA
Los Angeles, California

Background

A new rideshare service pilot seeks to establish best practices for public-private partnerships, featuring a flat-rate fare and employment opportunities for U.S. military veterans. In 2018, FASTLinkDTLA, the Downtown Los Angeles' (DTLA) transportation management-focused public-private partnership, was looking to monitor the need for on-demand accessible mobility options and began exploring different options for accessible mobility services that offered both flat-rate fares and employment opportunities (Figure 6.1). Gaps were identified in the transportation network through a few key indicators: increased demand for accessible options due to an underserved and aging population; the unique needs of female transit users such as off-peak hours and required cargo space, and a lack of first-mile and last-mile service options to the bus lines. Existing options for getting around included microtransit, such as dockless scooter services; express and managed lanes on highways; and bus services connecting different areas of the City of Los Angeles (LA). Though there are a variety of mobility options available, service is often not comprehensive enough in terms of operational hours or distance between destinations and service stops. Offering a solution to these gaps in the system and to serve more Angelenos, was FlexLA, a pilot program powered by moovel, a transit engagement platform set to change the ways people move in cities and partnered with ButterFLi, one of the City's first on-demand WAV services (Figure 6.2).

Women Driven Mobility: Rethinking the Way the World Moves   97

FIGURE 6.1 Marketing image for FlexLA showcasing the app and on-demand shuttles.

FIGURE 6.2 An accessible shuttle is pictured with a user and driver in view.

Through the FlexLA microtransit system, FASTlinkDTLA wanted to develop working solutions to the following questions that were realized during the data collection:

- How can transit serve as a solution to meet the needs of people who do not own a vehicle and are interested in moving downtown?
- How can DTLA meet the needs of women transit users to improve their comfortability with safety and access for things like cargo, strollers, and additional space they might not have access to on a public bus or micromobility services like bikeshare.
- Rideshare is not a true solution due to LA traffic trends. How can we offset congestion with more transit options?
- For visitors, how can DTLA reimagine micromobility and communications so that personal vehicles can only be parked once rather than being moved as the visitors access multiple destinations?
- How can DTLA lead in smart parking options with EV charging stations and AV options?

When looking at why traffic congestion and transit options are especially important for FASTLinkDTLA, it is important to consider LA's notorious reputation for traffic congestion. According to the FHWA, traffic congestion costs each LA driver an annual average of $1,774 in the form of wasted time and fuel. An estimated 82 hours per driver per year are lost to commuting between areas in LA. This congestion chokes both commuters and commerce and amounts to approximately $29.1 billion dollars in lost time and wasted fuel.

Goals and Objectives

With the FlexLA pilot, FASTLinkDTLA aimed to produce a cultural change toward new transportation options and was not positioned as a war on cars. The main goal was to reduce single-occupancy trips by 75% by 2030 by encouraging drivers in Los Angeles to use their cars differently and inviting visitors to experience the unique and customized transportation experience offered by FlexLA.

The objectives of these efforts aimed for these results:

- On-demand shuttle services that could fill gaps with expanded accessibility targeting women who might have different needs such as strollers, wheelchairs, young children, or cargo.
- Creating safer neighborhoods and communities by putting more eyes on the streets. Fewer people in personally owned vehicles can result in an increased walkability and create a culture of walkable spaces.
- Offsetting congestion with fewer vehicles on the road during peak times (5 p.m.-7 p.m.), pushing for Happy Hour, not Rush Hour.
- Economic development from people staying downtown longer after business hours.
- Identify and nurture opportunities to restore the tourism industry.
- Increase job opportunities for drivers.

Many workers and residents in Downtown L.A. are left with few transportation options during the late-night hours when public transit service is less frequent. FlexLA fills a gap in the city's transportation network, offering Angelenos a safe, reliable transportation option that connects with public transit and provides a more predictable cost than private rideshare services. FASTLinkDTLA will actively seek feedback from riders and the community to grow and evolve the FlexLA service to ensure it is meeting the needs of people who live, work, and visit Downtown L.A.

—Fixing Angelenos Stuck in Traffic (FAST) Executive Director, Hilary Norton

Approach and Challenges

Starting as a beta program in 2018, FlexLA offered Angelenos and visitors a flat-rate fare ($2) for point-to-point and on-demand service in energy-efficient vehicles. The vehicles of choice were Mercedes-Benz Metris passenger vans and GLC hybrid vehicles driven by U.S. military veterans. Using an app, users can hail a ride between the hours of 4 p.m. and 2 a.m. This app offers another option for users who take transit downtown or park once (allowing them to move between destinations without unparking their vehicle and parking it again—saving time and money and reducing emissions).

In addition to arriving at destinations or at their parking location, riders can also be picked up and dropped off at Downtown DASH bus stops. Rides can be shared by groups or by multiple users in a carpool functionality. FASTLinkDTLA partnered with ButterFLi, an organization that provides assisted rider specialty fleet vehicles, to offer additional vehicle options to groups of riders such as those requiring WAVs. Low-income users are able to register to ride for free. The app is offered in both English and Spanish (Figure 6.3).

FIGURE 6.3 Visualization of the application as it would be seen on an iPhone.

FIGURE 6.4 Sample map of the FlexLA service area in Downtown LA.

FlexLA is operated by moovel with fleet operator SMS Transportation. The FASTLinkDTLA microtransit service was led by the Executive Director of FAST in partnership with LA County Metropolitan Transportation Authority (Metro), the LA Department of Transportation (LADOT), LA Cleantech Incubator, and LA City Council Member Jose Huizar.

Funding for the FASTLinkDTLA project was provided through the managed lanes in LA County. The Metro ExpressLanes is designed to improve traffic flow and provide enhanced travel options for users. Anyone starting a FasTrak transponder account is charged a $40–$50 startup fee, which is loaded to the account as a balance; solo drivers are charged between $0.25 and $1.40 per mile; toll rates increase as traffic increases through congestion pricing. Carpool drivers are not charged a toll to use the managed lanes. These lanes dynamically switch to HOV-only to ensure a minimum average speed of 45 mph. LADOT sponsored this project and was responsible for overseeing the service delivery and connectivity with the local DASH bus system and regional public transit network.

Metro ExpressLanes and Silver Line also partnered with FASTLinkDTLA to provide first-mile and last-mile connectivity to the Silver Line Bus Route, one of LA Metro's most successful bus rules (Figure 6.4).

Results

LA is the first city in North America to pilot moovel's on-demand technology. The service "moovel" was designed as supplemental transportation to fill in transportation "blank spots." In its initial 2018–2020 pilot, moovel provided service to more than 2,000 passengers in six months. In the eight months of the pilot operation, FlexLA served more than 3,000 rides in Downtown LA with more than 14,000 downloads of the app. The FASTLinkDTLA pilot ran from October 2, 2018, to August 21, 2020.

FASTLinkDTLA is working with Metro and the City of LA to expand this pilot with a variety of new micromobility options and ButterFLi in the future.

About the Team

FASTLinkDTLA is a partnership of businesses, employers, developers, business improvement districts, organizations, and nonprofits focused on improving mobility, transportation, and infrastructure in Downtown LA. FASTLinkDTLA's activities include advocating and promoting transportation system and demand-management strategies to reduce traffic congestion, air pollution, and commuting costs. It also generates joint public-private partnerships to solve transportation problems. In addition, FASTLinkDTLA creates a central information service for increasing mobility choices and addresses public transportation and other transportation-related issues with the goal of improving quality of life.

Hilary Norton brings more than 28 years of experience in transportation and community development to her commission role.

As FAST's founding Executive Director since 2008, Norton has mobilized a diverse coalition of business, labor, and civic groups, educational institutions, and transit organizations to support policy and infrastructure improvements to LA's mobility, livability, and economic prosperity. FAST's major initiatives include FASTLinkDTLA, a new Transportation Management Organization for Downtown LA that operates one of the first-ever LA County microtransit systems to connect travelers through the FlexLA multimobility app and connects new on-demand WAVs, transit, vanpools, carpools, scooters, bikeshare, biking, and walking options. Mobility hubs such as carshare, bikeshare, bike parking, EV charging, and traveler services at transit stations, job, and education centers; comprehensive arterial improvements to improve travel time, encourage mode shift, and promote safety and transit connectivity; Metro ExpressLanes implementation throughout LA County, creating the Metro ExpressLanes Business Roundtable to support the pilot corridors and expand into a network; expanding LA County's BRT network; and the Sixth Street Viaduct, Sixth Street Park, and Arts District Station, which is the largest bridge reconstruction project in LA's history, adding bicycle and pedestrian lanes and connections to the LA River and the Metro Red/Purple Line.

Norton served as 2018 chairwoman of the LA County Business Federation (BizFed) and is on the Board of Directors of the Central City Association. She co-chairs the Transportation Committees for BizFed, the LA Chamber of Commerce, and the LA Business Council.

Norton also served as the Business Representative on LA County Metro Policy Advisory Committee and was a member of the Advisory Boards for Metro's Office of Extraordinary Innovation and Metro's NextGen Bus Study. She is a member of SCAG's GLUE Council and its TDM Working Group. Norton is a Director for the Orthopaedic Institute for Children, which provides world-class health care to all children regardless of ability to pay, and is a Board member of the Leo Buscaglia Foundation.

She holds a B.A. from Wellesley College and a Master of Public Policy from Harvard University's John F. Kennedy School of Government.

7

Placemaking

As more women gain influence in the community planning industry, we will see a shift in priorities. This shift can't come soon enough, as we move from a fossil fuel driven transportation system that exacerbates climate change to a more holistic approach that includes not only electric vehicles—and the vast network of charging stations necessary to power them—but reliable, connected, and accessible public transit systems that move people cleanly and efficiently. Women frequently understand more deeply the underlying inequities—for women, for people of color, for the disabled—embedded in our auto-dominated transportation systems, and are emboldened to advance the innovations and policy change that both improve lives and increase prosperity, while reducing the harmful environmental impacts our current systems impose on our Earth.

—Andrea Brown, AICP, Executive Director, Michigan Association of Planning

The profound words of Jane Jacobs, urbanism's driving force, are one of the first things you might hear in planning school. Her nontraditional journey from journalism to urban planning, her flowery language that shared the truths—good and bad—about places, and her unrelenting love for cities echoes for the reader long after closing the *Life and Death of Great American Cities*. The book, published in 1961, is the source of the chapter-opening quote, which somehow still does not feel like it has been fully applied and offers reference to something that us city lovers are still aspiring to half a century later.

Planners have been talking about equity in cities and communities for more than 50 years, and we are still looking for it and applauding it when it arrives. For Jacobs, and for the many urbanists who followed in her footsteps, equity and inclusive place-making was never seen as something that should be an optional aspiration, but rather was the baseline. Without the consideration of all community members, cities end up with lackluster design, insufficient facilities, and barriers for certain populations. Though there are many populations to consider when planning places, there is no question that cities, as we know and love them today, were not built for women. Cities were designed for and shaped by men.

These men continue to literally write the book on urban planning, and though they represent different facets of the practice, most well-respected and popular planners only represent one demographic group: white men. In 2017, Planetizen created the list of the 100 Most Influential Urbanists of All Time, going back to 498 BCE. Of this list, 18 are women and only three of them are women of color—totaling a dismal 18%. Granted, this was not a scholarly article focused on metrics or comparing works, but it shows just how little women are recognized in the planning world, though they have always been a major part in the way cities exist.

In the profession, there is a long-running trend of planning conferences and talks and hosting all-male speaking panels, or "manels." Though not officially recognized as such, men take the stage, while the women participate from the audience. This can be seen in cities, too. While it is refreshing to see a growth in women in leadership in the political arena, C-suite offices, and representing these places as staff, there is still a demand for the female perspective on cities that needs more attention. According to the U.S. Census Bureau ACS Public Use Microdata Sample, nearly 60% of the urban and regional planner workforce is male, and 75% of both genders in this workforce is white. The lack of diversity represented in the profession is not only alarming, it is not representative. While this lack of representation is problematic, there is good news: the planning profession is growing rapidly, meaning more places are placing a value on the practice and allowing for intentional growth in the underrepresented gender and race categories. Between 2018 and 2019, the number of people employed as urban and regional planners had grown 92.6% from 20,051 to 38,620. There is currently a gender wage gap of approximately $19,000 in the profession.

This divide in both the built environment and in our classrooms, boardrooms, and conference centers is a large threat to our communities because this lack of representation applies to about one-half of the populations that move through these places.

There is an especially large discrepancy between the ways men and women move about. In various studies, men are shown to take more direct, faster trips and have increased access to multiple modes of transportation. Women have very different needs on their trips and face additional barriers. What makes a city safe or enjoyable has been considered from the perspective of an able-bodied person with two free hands—in daylight. Women often have full hands, maybe pushing strollers, keeping one earbud out while listening to their favorite songs, and immediately identify the exit when entering public places. Rethinking our cities for and with women and for all genders is not simply a matter of equality, it is a matter of safety, and it is necessary.

In this chapter, we will visit Detroit, Michigan, and Minneapolis, Minnesota. We can reimagine the built environment by breathing new life into existing infrastructure, designing streets for people instead of cars, and thinking about the long haul—not just how people get from Point A to Point B, but how they get to Point Z, enjoying every stop along the way.

In the great, loud words of Jacobs, "Cities have the capability of providing something for everybody, only because, and only when, they are created by everybody."[1]

Building a Vibrant Urban Innovation District

Mary Culler, President, Ford Motor Company Fund and Detroit Development Director, Ford's Michigan Central project

Carolina Pluszczynski, Detroit Development Manager, Ford's Michigan Central project

Julie Roscini, External Engagement Manager, Ford's Michigan Central project

Detroit, Michigan

---

[1] *The Death and Life of Great American Cities*, p. 238

Background

We stand at the precipice of the biggest shift in transportation in more than a century. Technologies like connectivity, automation, electrification, artificial intelligence, and the sharing economy are dramatically changing the automotive and mobility landscape. At the same time, through congestion, pollution, and equity issues, our communities, cities, and the planet are facing new problems that require new solutions. With its sights set on the future, Ford Motor Company wants to be a leader in this revolution, not just another player.

Born out of disruption, Ford has always looked at innovation in a different way. As they reimagine the industry for a new century, the automaker has turned to Detroit—the birthplace of mobility—to be the epicenter and culmination of their vision for the future.

In January 1988, the last train left Michigan Central Station, the historic former intercity passenger rail depot in Detroit, Michigan. For 30 years, the city's iconic station sat abandoned and derelict—viewed as a symbol for Detroit's fall from greatness. Then, in 2018, Ford purchased it as the crowning jewel of Michigan Central, a new mobility innovation district the company is building in Corktown, Detroit's oldest neighborhood.

For Ford, Michigan Central is about preparing the company, Detroit, and the State of Michigan for another century of innovation and success by designing new urban solutions for the way people and goods will move around in the future. The goal is to attract innovators from the mobility space and beyond who want to come together to solve complex problems and create something truly unique that brings renewed opportunities, sustainability, and vitality to the area.

Detroit led the biggest mobility revolution in modern history, and now Ford is creating a physical space surrounding Michigan Central Station to act as the launchpad for what comes next. No place in the world is more important for transportation's past, present, and future.

Goals and Objectives

Ford is investing more than $740 million in the development of the new Detroit campus, including the restoration of Michigan Central Station, one of the city's greatest historic assets, which will be the centerpiece of the new tech hub. When completed, it will be an open platform for innovators, startups, entrepreneurs, and other partners from around the world to develop, test, and launch new mobility solutions on real-world streets, in real-world situations. Michigan Central will also be a node on the State's proposed 40-mile connected and AV corridor project, led by Cavnue—a company founded by Sidewalk Infrastructure Partners—to build the future of roads between Detroit and the suburb of Ann Arbor, linking the district to a broader regional network of automotive testing, research, and innovation.

From the beginning of this project, Ford established five guiding principles that drive strategic decision-making:

- Build the Future of Mobility: Showcase new technologies that are flexible, accessible, and provide options for better transportation.

- Optimize for Innovation: Promote serendipitous interactions between people, businesses, and ideas.

- Make an Inclusive and Authentic Place: Create a place that is authentically Detroit, invites in the community, and supports diverse and equitable uses.

- Celebrate Heritage and Legacy: Preserve the area's unique character and historic structures while introducing new forms that respect local heritage.

- Make Decisions That Create Equity: Create partnerships that bring opportunities, environmental sustainability, and vitality to the area.

Approach and Challenges

Ford is building the future of mobility in a place that is authentically Detroit. Michigan Central will be an inclusive, vibrant, and walkable community made up of new and revitalized buildings, open spaces, a first-of-its-kind mobility testing platform, and 1.2 million square feet of commercial space. It will not be a Ford campus or a traditional real estate development but rather a dynamic innovation ecosystem that is open to everyone and connected to the neighborhoods and city around it, the region, and the world beyond. Michigan Central Station will be open to the public with locally inspired restaurants, shops, hospitality, arts and cultural activities, and public amenities, in addition to modern office spaces. It will be a must-see destination to experience the best that Detroit has to offer.

Existing and new forms of mobility will become part of the streetscape around Michigan Central—sharing the roads with pedestrians. Ford is transforming an

abandoned area of old railroad tracks to the south of the station into a first-of-its-kind mobility-testing platform. This space will be an open, versatile landscape where Ford and other partners can test and showcase emerging technology to move people and goods, including AVs, electric bikes (e-bikes), scooters, and other micromobility initiatives and last-mile solutions. The mobility platform will also provide shared paths for pedestrians and cyclists and gathering spaces for the community that are reconfigurable for a variety of uses.

The Book Depository, another Ford-owned property that sits adjacent to Michigan Central Station, will be one of the first buildings in the district to come online and is designed to support flexible, hybrid work. The interiors will be highly adaptable and versatile so that anything from walls and panels to furniture and fixtures can be flipped, moved, or reused to support multiple uses. From conferencing to strategy rooms, all spaces will be tech-enabled to allow for real-time digital and virtual working among remote team members and collaborators.

The buildings and open spaces within Michigan Central are being designed to create team adjacencies and foster environments in which diverse groups of people can innovate, collaborate, and interact with each other and the world around them. In addition to the 2,500 Ford workers who will be based there, Michigan Central will also bring in 2,500 additional innovation partners and tenants—from large corporations to startups, entrepreneurs, researchers, venture capitalists, civic organizations, even competitors—to work side by side with local residents, businesses, and artists.

An added complexity is that Ford will be moving into an existing neighborhood, one with a rich history and culture. The company is committed to being a good neighbor and wants Michigan Central to become part of the fabric of the neighborhood. The local community is an integral part of the vision for the mobility innovation district, and having their voices heard and incorporated has been essential. Since the project was first announced, Ford has continued to work closely with community members to listen to their needs and concerns through one-on-one conversations, community meetings, neighborhood newsletters, and listen-and-learning sessions, and by supporting local events and activations.

Next Steps

Ford is taking a phased approach to this development. It already has 250 members of its AV team based at the Factory, a refurbished old hosiery factory on Michigan Avenue, and they have been testing self-driving vehicles in Detroit since June 2019.

Work on the Book Depository and the nearby Bagley Parking Hub began in the first quarter of 2021, with both buildings expected to open in early 2022. Michigan Central Station, currently in the middle of phase two of its restoration, will be complete by the end of 2022, and the first tenants are expected to take occupancy in 2023.

In undertaking this enormous project, Ford is making a long-term commitment to Detroit. It wants to make a positive impact on the neighborhoods around Michigan Central and the city at large. The automaker's hope is that Michigan Central will be pivotal not just for the company but for the future of Detroit and the future of Michigan—ensuring it remains at the forefront of mobility innovation and automotive technology.

Ford is returning Michigan Central Station to its original grandeur to become a welcoming place for all Detroiters. It will become a must-see destination for the community and for the City of Detroit. As has been shown by redevelopment projects across the country, a major catalytic project such as this helps spur additional development around the anchor developments and will help bring about a larger resurgence of the surrounding neighborhoods and region. This project will help boost economic growth, create new jobs, and attract world-class technical talent to Detroit and Michigan.

About the Team

Ford's Michigan Central placemaking project is a large effort for the century-old manufacturer, with three key women at the helm: Mary Culler, Carolina Pluszczynski, and Julie Roscini.

Mary Culler is president of Ford Fund, a role she assumed January 1, 2020, overseeing the company's philanthropic investments through signature programs and employee volunteer outreach in more than 50 countries. In addition, Culler is Chief of Staff, Office of the Executive Chairman, reporting to Bill Ford. She has responsibility for advancing strategic priorities and shaping internal and external engagement in areas such as sustainability, smart mobility, autonomous driving, and corporate citizenship. She is also Development Director for Ford's Michigan Central Station project, overseeing the strategic direction of the historic train station redevelopment and other Ford properties in Corktown.

Prior to her current positions, Culler was Director, Ford U.S. State and Local Government Relations. In this role, she managed Ford's engagement with policymakers nationwide on a wide range of automotive and community issues. Her team also negotiated investment and job creation incentives in Ford plant states.

Her other positions at Ford included managing Ford's North Central Region's Government Relations Office in Chicago, Illinois, where she spearheaded the company's Operation Better World community and dealer relations efforts. She led Ford's Global Public Policy office in Dearborn, Michigan, managing the development of key company policies and positions.

Prior to joining Ford in 1999, Culler worked for Chicago Mayor Richard M. Daley as manager of the city's Industrial Development Program. In that capacity, she worked with businesses on economic development and job creation initiatives.

Culler also held several positions at the U.S. Environmental Protection Agency in Washington, DC, including a role as Deputy Director of the National Brownfields program where she launched the national program to remove regulatory barriers regarding the cleanup of abandoned industrial properties. She also worked in the U.S. Senate on energy and natural resource issues for the Chair of the Energy and Natural Resource Committee.

Culler holds a bachelor's degree from Indiana University and a master's degree in public administration from Harvard University. She lives in Birmingham, Michigan, with her husband, Andy Norman, and their three children.

Carolina Pluszczynski is Detroit Development Manager responsible for overseeing strategy and delivery for Ford Motor Company's Michigan Central development in Corktown, Detroit's oldest neighborhood. Prior to this role, Pluszcynski was part of Ford's Transformation Office, where she led company-wide initiatives accelerating changes in corporate governance, implementing product-driven teams, and progressing diversity and inclusion.

Throughout her 20-year career with Ford, she has taken on several leadership roles, including as Chief of Staff to the Mobility President and in the CIO's Office, where she spearheaded employee engagement strategies and communications for the company's global transformation of IT. Pluszczynski has managed technology programs in manufacturing, product development, and Ford customer service, including vehicle scheduling systems, CAD/CAM software releases, and business IT. She joined Ford in 1999. Prior to that, she ran her own consulting business focusing on technology projects.

Julie Roscini is the external engagement manager for Ford Motor Company's Michigan Central development in Corktown, Detroit's oldest neighborhood. In this role, she is responsible for creating a partner ecosystem for Michigan Central's mobility innovation district. She also leads external programming for the development, engaging the community and future partners as Ford builds out its vision for Michigan Central. Roscini joined the company in 2010 and spent most of her career in IT and mobility, working on communications, strategy, and planning. She earned a bachelor's degree from Wayne State University and a master's degree from the University of Michigan.

Development without Displacement: Creating 20-Minute Neighborhoods

Janet Attarian, Principal and Senior Mobility Strategist at SmithGroup; formerly Deputy Director, City of Detroit Planning and Development
Detroit, Michigan

Background

In 2013, Detroit filed for bankruptcy—the largest municipal bankruptcy filing in U.S. history by debt.[2] The city had amassed an estimated $18–20 billion in debt, and its population had declined from a peak of 1.8 million citizens in 1950 to only 700,000 in 2013.

Thirteen months later, Detroit emerged from the bankruptcy, and private investment in the downtown area was on the rise. However, beyond the immediate downtown area, there were still struggling neighborhoods, and in fact, much of the economic resurgence was not penetrating across the other 143 square miles within the city limits. There was still a lot of work to be done.

Following the bankruptcy, the city organized a new Planning and Development Department with the mission to revitalize the underserved neighborhoods. For decades very little to no investment had been made in the city's neighborhoods. Starting with targeting areas of strength throughout the city, and acting as a catalyst to drive additional investment and development to improve Detroiters' quality of life,

---

[2] https://en.wikipedia.org/wiki/Detroit_bankruptcy

the new department was given the task of helping communities and building trust in city planning.

The City of Detroit Planning and Development Department's vision cited building a healthy and beautiful Detroit, built on inclusionary growth, economic opportunity, and an atmosphere of trust. The secret to achieving this vision, the Department believed, was based on the simple but powerful idea of making Detroit walkable.

Once the initial challenges of identifying the neighborhood planning areas and getting the first three plans funded was accomplished, the pivotal challenge was to figure out how to fund and create the internal capacity to implement. the projects that were being identified through the planning processes. The city had no capital budget and had not done any capital planning for years, and its bond rating was poor. Private development without the city's financial assistance was all but impossible in Detroit's neighborhoods. Neighborhood infrastructure improvements had been minimal for decades and were almost exclusively focused on basic maintenance. The city's implementing departments had little internal infrastructure or resources in place to execute the visions that were coming out of the Planning Department even if funding was identified, and there were almost no city programs for doing so.

Goals and Objectives

The key to revitalizing Detroit's neighborhoods and making a walkable Detroit would be the ability to improve streets and sidewalks and to create a complete streets program to do it. Complete streets is a transportation policy and design approach that requires streets to be planned, designed, operated, and maintained to enable safe, convenient, and comfortable travel and access for users of all ages and abilities regardless of their mode of transportation.[3]

---

[3] Ritter, J., "'Complete Streets' Program Gives More Room for Pedestrians, Cyclists," *USA Today*, July 29, 2007, Retrieved August 23, 2008.

The desired outcome of the neighborhood plans would be a series of walkable nodes, or "20-minute neighborhoods" where Detroiters could conveniently and safely access their daily services and needs within a 20-minute bicycle/micromobility ride or walk of where they lived. The goal was to create growth and development without displacement by:

- Fostering vibrant, growing neighborhoods across the city.
- Preserving all regulated affordable housing.
- Ensuring that wherever growth occurred, it increased inclusion and reduced segregation.

For Detroit's streets in particular, the city implemented a set of objectives for sustainable mobility in Detroit, which included:

- More affordable and equitable multimodal choices and better access to transit.
- Improved safety, especially for pedestrians, bicyclists, transit riders, and other vulnerable users.
- People-centered streets, with a sense of place, that are attractive and support local small businesses.
- Increase access to healthy forms of mobility and safe places to walk.
- Resilient, restorative transportation infrastructure.

Approach and Challenges

The first challenge was to see if Detroiters agreed with the set objectives and to determine which of the goals aligned with the needs of each of the neighborhood areas surveyed.

Trust was already an issue, and new ideas are difficult to discuss when trust is absent. To overcome this obstacle, the department developed and tested a wide range of engagement strategies, which included going out into the community and listening, engaging with citizens, and educating on the tools we had as physical planners. This would help to gain trust.

In the first planning area on Detroit's northwest side, the department hosted more than 50 public meetings to help define what could and should be done in the area. This concluded in a community vote with the Mayor in attendance nearly three years after the process began. Not only were there meetings large and small, but they also used tactical urbanism to pilot projects so that both constituents and the city departments could test drive concepts and debate new ideas such as two-way cycle tracks and the reduction of road lanes. Tactical urbanism includes low-cost, temporary changes to the built environment, usually in cities, intended to improve local neighborhoods and city gathering places. Pilots included pop-up protected bike lanes, community-painted walkways, and temporary placemaking activations.

As time went on, the department added more engagement strategies, including working with youth and having them present their ideas at public meetings, community bike rides, hiring members of the community to help go house to house and conduct surveys, hosting all-day open houses, and providing more pop-ups, such as creating a plaza out of part of a three-way intersection along a major corridor. From this work, the department saw that the communities choose to prioritize walkable streets, expanding sidewalks, adding bike lanes, slowing dangerous traffic speeds, and adding amenities in their communities.

The second challenge was to fund the identified projects. As mentioned earlier, there was little to no city funding for this work, and given the very limited and competitive nature of state funds through the Transportation Alternatives Program, the millions of dollars needed to transform Detroit's streets was going to be difficult to come by. To solve this problem, the department worked with the city's finance office and Public Works Department to create the first city bond since before the bankruptcy. Given the city's bond rating at the time, this was no easy task, but the State of Michigan DOT was planning to give Detroit a modest increase in annual ACT-51 funds for roadway improvements, and instead of having this small increase have a minor impact over many years, the department helped craft a bond program, which was approved by the Mayor's office and City Council, to issue a $125 million bond based on this additional revenue. This provided the financial security that allowed the city to get a good interest rate and a good rating on the bond. The city financed an $80 million place-based complete streets program out of the $125 million, with an additional $26 million for road rehabilitation projects and $20 million for sidewalk reconstruction. After many levels of review, the whole package was approved.

The final challenge was to create the internal capacity to get the work done. The city had no transportation planners and no complete streets program and had not done any streetscape work through the city itself in a long time. The department started this process by securing a small amount of funds to provide a workshop for internal staff on how to create and implement a program. This led to securing funding for hiring the city's first transportation planning team, a dynamic group of women who continue to lead the city's efforts to this day.

Results

These efforts have had a lasting impact on the City of Detroit. First and foremost, several of the street projects funded through the bond have already been built and brought to reality, and several more are under design or construction. These redesigned streets included protected and raised cycle tracks, road diets, shared streets, and in-lane bus-boarding, along with wider sidewalks, better ADA access, improved safety, trees, amenities, and green stormwater infrastructure—all things the department had laid out in their original vision when putting the projects together for the bond and working with her teams as they planned with communities.

The Department of Public Works created a complete streets program, bringing on board the entire transportation team to get these projects designed and implemented. Throughout the project, it was determined that many of Detroit's important neighborhood streets were under state jurisdiction rather than city jurisdiction. The department worked closely with the Mayor's office and Governor's office to align the city's work with the state's work and was instrumental in getting millions of dollars in state funds to match the city's funds, allowing the completion of a strategic corridor, Grand River, in record time.

As part of the bond effort, the department had to show how it was part of a larger funding strategy for the city's roads by working closely with the Department of Public Works and creating a five-year capital plan. Even more important, this was part of a larger effort to create a capital plan for the city that began the process of identifying the funding that would be required for other aspects of the city's planning efforts.

Perhaps, most important, with the funding of the bond and the $80 million of place-based street projects that resulted, the city was able to put a down payment into the Strategic Neighborhood Fund. The city demonstrated that it was bringing money to the table when it asked private businesses and nonprofits to donate to the fund to create a meaningful funding stream for all of the initiatives coming out of the department's neighborhood planning efforts.

About the Team

Janet L. Attarian, AIA, LEED AP BD+C, has more than 25 years of experience in creating beautiful, livable cities with a focus on inclusive neighborhood placemaking, mobility planning, and street design. In her leadership role as Senior Mobility Strategist, she helps craft SmithGroup's vision for transportation planning and mobility policy that is focused on people and the planet. She has a gift for synthesizing the multiple disciplines required to create vibrant streets and communities and to drive innovative policy. She brings her years of experience implementing complete streets, creating healthy neighborhoods, and building sustainable infrastructure to drive innovative, dynamic plans and projects that help create resilient, equitable futures for cities.

Before joining SmithGroup, Attarian was Complete Streets Director for the City of Chicago's DOT, including the city's Bicycle, Pedestrian, and Streetscape Programs. While there, she led numerous policy efforts, including the Department's Complete Streets: Design Guidelines, Sustainable Urban Infrastructure Guidelines, Streetscape Design Guidelines, and Make Way for People programs. She led the Bloomingdale Trail and Park Framework Plan, Millennium Park Bicycle Station, and Navy Pier Flyover. She also led the planning, designing, implementation, and maintenance of more than 100 Complete Streets projects including the reconstruction of Congress Parkway and Fulton Market Shared Street. She helped design and implement public plazas throughout the city and activate them with pop-up retail and art, including innovative public-private partnerships. She continued this work as Deputy Director of the City of Detroit's Planning and Development Department, where she helped lead a series of inclusionary neighborhood planning frameworks that reimagined the city's streets by driving place-based mobility projects, including two-way cycle tracks, streetscapes, and new plazas, applying an equity lens to neighborhood planning and implementation. Attarian also help lead the development of the city's Strategic Plan for Transportation and drove the development of a five-year capital plan for the Detroit Department of Public Works, including a three-year capital bond for the design and implementation of walkable retail corridors in 15 strategic disinvested neighborhoods.

Attarian speaks around the country on complete streets, sustainable infrastructure, and equity, and her work has been featured in numerous publications including the *New York Times* and the FHWA's *Public Roads* magazine. She currently serves as a member of the FHWA's Sustainable Pavement Technical Working Group, an advisor to Remix, and is a member of the Board of Directors for MoGo, Detroit's bikeshare system. Prior to working with the City of Chicago, Attarian had her own design firm and worked for several architectural firms including DLK Civic Design. She has been a licensed architect since 1996 and has a master's degree from the University of Michigan.

Designing Community-First Transit, 2010-2018

Katie Walker, Director of Innovation and Research at the Minnesota DOT, formerly Director of Southwest LRT Community Works
Hennepin County, Minnesota

Background

In May 2010, the Southwest Light Rail Transit (LRT) project's locally preferred alternative (LPA) was chosen by Minnesota's Metropolitan Council and received FTA approval to enter preliminary engineering the designs would be developed with interagency support and significant community input. The Southwest LRT Project was proposed as a 15-mile LRT line serving the Twin Cities southwest region, five municipalities with 17 stations in Eden Prairie, Edina, Hopkins, Minneapolis, Minnetonka, and St. Louis Park. This LRT line would increase the existing system's capacity in response to a high demand in the service areas and would provide an alternate travel option for choice riders to benefit transit-dependent populations. In addition to the new line itself, the project would provide further regional connectivity between the Central (Green Line) LRT and Hiawatha (Blue Line) LRT lines as well as the Northstar commuter rail line and the Metro Transit bus system.

Historically, the design/development for major transportation infrastructure projects are done in isolation from adjacent land use, leading to lost opportunities. The Southwest LRT Community Works project was initiated with the intention of creating a new model for transitway development through systemic organizational change to institutionalize consideration of the following:

- Affordable housing.
- Economic development.
- Economic access.
- Educational access.
- Food access.
- Land use.

- Service and retail access.
- Workforce development.

Goals and Objectives

The Hennepin County Community Works approach was used for this project. Hennepin County Community Works focuses on partnering with cities and other agencies, businesses, neighborhood organizations, and county residents to build the long-term value of communities; create and sustain great places; and make quality investments in redevelopment, transportation, public works infrastructure, parks, trails, and the environment.

The goals of the Southwest LRT Corridor Community Works program were to:

1. Promote collaboration across disciplines, agencies, and communities.
2. Realize and communicate shared goals of delivering an on-time, on-budget LRT project.
3. Encourage transit-friendly and pedestrian-friendly development.
4. Leverage private resources to augment public resources for a more comprehensive system.
5. Improve the quality of life for nearby communities and the region.

Although LRTs are generally characterized as a "jobs line," a closer look at the line and its stations revealed that it is much more. The Southwest LRT Corridor is an interwoven string of interesting and unique places. While each station is distinct, there are a number of systems and, in some cases, common elements, features, and characteristics that help connect station areas along the corridor and relate them to adjacent neighborhoods. When it opens, the Southwest LRT will connect a range of diverse communities and place types along its route including places to work, mixed-use centers, public institutions, and important recreational and open-space assets.

Approach and Challenges

In October 2010, the Twin Cities region was awarded a U.S. Department of Housing and Urban Development (HUD) Sustainable Communities Planning Grant intended to build on existing regional planning efforts to advance multimodal transportation choice, promote affordable housing with access to jobs, foster transit- and pedestrian-friendly development, support environmental preservation, and increase energy efficiency.

The grant funds were targeted to support comprehensive transit corridor plans that include strategies to:

- Provide access to living-wage jobs.
- Provide affordable and life-cycle housing choices.
- Align workforce opportunities with corridor employment prospects.
- Improve connections to sources of fresh locally grown and ethnic foods.

In addition, the effort was intended to build local implementation capacity within regional corridors to advance interjurisdictional planning, regulatory, and administrative efforts that will support the Building Sustainable Communities program in seven Twin Cities agencies that were introduced by LISC.

A new process was used to concurrently design the Southwest LRT project and identify station area needs for the half-mile mile area around each proposed station. This was a 14-month process that included nine months of intensely focused efforts to have an open, transparent, inclusive, and iterative discussion of how modifications to the placement of the LRT infrastructure (e.g., tracks, station platforms, park-and-ride facilities, etc.) could influence and affect the long-term development potential in the station areas. This was accomplished through close to daily meetings/conversations and idea exchanges between the Southwest LRT project office staff, the consulting team, and Southwest LRT Community Works partner agency staff. The intentional inclusion of community development staff in meetings with the engineering staff greatly facilitated this open dialogue and led to a successful process.

Results

A key result was the Southwest Investment Framework, which acts as a living document that guides the public and private sector investments necessary to facilitate the evolution of the LRT station areas into transit-oriented developments (TODs) with a unique sense of place that relates positively and synergistically with the Southwest Corridor as a whole.

The results of this process are action plans that will assist the cities, Hennepin County, other Community Works partners, the Southwest LRT project office, and the private sector in understanding infrastructure investments that are needed to improve business and housing conditions in the near term and to enable the station areas to achieve their long-term vision and the LRT project to increase its ridership base. The TSAAP process was also used as a tool to recommend changes to the LRT engineering to better serve existing land uses and facilitate long-term phased TOD. Identified more than $120 million in infrastructure (last-mile connections and utilities) within a half-mile of the LRT line to set the stage for future development.

The results of the intense collaborative process included significant changes to the LRT engineering that will result in a lower cost, higher ridership LRT line with greater potential for enhanced TOD.

About the Team

Katie Walker currently serves as the Director of Innovation and Research at the Minnesota Department of Transportation (MnDOT), where she is charged with creating a culture where research, innovation, and analysis bridge today with tomorrow to improve the quality of life for Minnesotans. Prior to joining MnDOT, Walker served in a variety of roles at state, regional, and local governments including stints as a planner for Metro Transit/Metropolitan Council, Dakota County, and the 494 Corridor Commission.

During her 20-year tenure with Hennepin County, Walker held a variety of roles, including Southwest LRT Project Manager, Southwest LRT Community Works Program Director, and Public Works Policy and Planning Manager. Most recently, Walker headed strategic initiatives for the county's Center of Innovation and Excellence. She holds a Master of Public Policy degree from the University of Minnesota's Humphrey School of Public Affairs, a Mini-MBA from the University of St. Thomas, and a bachelor's degree from the University of St. Thomas.

Policy and Legislation

Courtesy of Molly Korn.

Good designers know the best ideas are borne out of set parameters—a way of taming and focusing on the infinite, often overwhelming, possibilities. Parameters give direction, urgency, drive.

This is the role of good policy and legislation. Policy can chart a course, provide a direction, create opportunities and protections. Policy gave us the New Deal, freeways that took us from coast to coast, opportunity for all, the American Dream.

In this moment of new mobility, policy will set the parameters to chart our course to the future.

—Mallory McMorrow, Michigan State Senator, 13th District

The importance of policy on the transportation network cannot be understated. Everything made possible by state DOTs, federal agencies, and private entities rely on the ways that policy provides license to shape the social, economic, and physical landscapes. In consideration of the spaces where transportation systems exist, our railways, highways, and streets are all arranged based on land use and were historically developed for trade and commerce. Additionally, the form of neighborhoods is also carved out by transportation networks. Development patterns and the transportation systems that are shaped around them additionally determine social decisions such as where people choose to live or to work and, depending on the cost of living in areas directly adjacent to transportation, could provide populations the opportunity to participate in those communities—and prohibit others from the same opportunity.

Unfortunately, much existing transportation and planning policy has been discriminatory, exclusionary, or not comprehensive in nature and has done unintentional damage to communities and to the natural environment through forest removal, greenfield development, disinvestment, socioeconomic segregation, and physical removal of neighborhoods. The Federal Highway Act of 1956 stands out as a once-in-a-generation policy with 41,000 miles of highway. Though highways were initially designed to serve as a tool of national defense, they quickly began to make it easier for people to come and go from place to place. One of the largest and most important pieces of legislation to come as a result of this act was the NEPA, which was signed into law in 1970 and states that it is the national policy "to use all practicable means and measures, including financial and technical assistance, in a manner calculated to foster and promote the general welfare, to create and maintain conditions under which man and nature can exist in productive harmony, and fulfill the social, economic, and other requirements of present and future generations of Americans."

Following the construction of highways through federal investment, there was a level of realization that programs at such a high level can often cause unintended consequences. The NEPA was the first major environmental law to be adopted in the United States, and to this day, it sets the precedent for the ways that federal programs can be implemented.

In 2009, the American Automakers Policy Council (AAPC) was formed—spearheaded by Ford Motor Company, GM, and Stellantis—to represent the common public policy interests of its three member companies. The AAPC was created as a conscious and collaborative effort to elevate common automaker issues impacted by policy and to work together to advocate for overall industry support at a federal level. These common threats to automobile development, manufacturing, and sales include:

- Automotive standards: As the demand for U.S.-built vehicles continues to grow globally, acceptance of U.S. vehicle standards around the world is essential.
- Environmental stewardship: Showcases fuel efficiency and other green policy and vehicle advancements.
- International trade: Involves building a strong U.S. economy based on our manufacturing leadership.

- Job creation: Focuses on increased hiring of a new type of workforce that is efficient and technologically advanced.

- Research and development: Growing and developing an expertise in the scientific and technological advancements within the industry.

The policy and legislation needed in the United States go far beyond road infrastructure and automakers. Although transit can service more than just the personal vehicle, most of the existing policy in transportation is car focused. However, that is starting to change with the Biden-Harris administration, especially at the state level.

In the introduction of this book, Michigan Governor Gretchen Whitmer painted a picture of what government involvement can look like in advancing mobility for its citizens, from the creation of the Office of Future Mobility and Electrification, which focuses on economic growth and job creation, to the PlanetM Grants (Section 3.1) that allowed the state to be a recognized mobility test hub, globally.

This is just the beginning.

As the world continues to grow global commerce, we see rapid advances in technology with connected autonomous vehicles, and through smaller solutions like drones, our needs as a nation are changing, and it is critical that policy leads to offer quicker adoption for these new systems that are more economic, faster, and often have lower emissions. Transportation policy aims to make efficient and effective decisions on how resources for transport are allocated and managed while also continuing to regulate existing transportation networks. While there is some overlap between the ways in which public and private entities engage with these policies, it is our federal and state policies that we lean on the most to set the standards for how national transportation systems will be designed, constructed, operated, funded, and maintained.

In this chapter, we will look at some of the most innovative approaches to policy of the next generation, focusing on equity and cross-collaboration between sectors. Because everyone's transportation needs are different and vary in scale, distance, and mode, policy cannot be one size fits all. Diverse voices must lead the discussion on what policies include and exclude and what they mean for the industry as a whole.

Commission on the Future of Mobility

Alisyn Malek, Executive Director, Commission on the Future of Mobility
Washington, DC

Background

Opening the first-ever Global Sustainable Transport Conference in November 2016, UN Secretary-General Ban Ki-moon told delegates gathered in Ashgabat, Turkmenistan, that the world has the resolve, commitment, imagination, and creativity "to transform our transport systems in a sustainable manner that will improve human wellbeing, enhance social progress and protect our planet."

The former UN Secretary-General was correct. Transportation transcends economics. It has a human side, and we should all be concerned about policies that do not put people first and provide the safe, affordable, and sustainable access they deserve.

Fast forward five years, and this sentiment around setting a new direction for global transportation mobility policy that connects people and communities to jobs, goods, and services, and leaves no one behind is still essential to our long-term global prosperity and perhaps even more opportune.

Today, emerging technologies such as AI, machine learning, big data analytics, and the IoT are accelerating new businesses and service models across the transportation sector's efforts to move people and goods around the world.

But in cities from LA to Paris to Delhi, many people are still left without safe, environmentally friendly, and convenient options to get from Point A to Point B. Meanwhile, technology threatens economic opportunity for many as society scrambles to prepare for a workforce transition that is already underway while much of our physical infrastructure sits close to the breaking point. Add in a growing global climate crisis and the need to significantly reduce carbon emissions in the transportation sector over the next half-century, which has become a formidable but necessary task.

In parallel, the challenge persists of working with governments that are traditionally slow and deliberate in crafting regulations that tend to remain in place and unchanged for long periods of time. They also often work within legacy frameworks and regulatory silos to create or modify policies, enforce them, and communicate them to the public at a previously undreamed-of pace.

Today's mobility paradigm demands a new global, forward-thinking approach to understand how we got here and develop new policies that will provide for investment, address environmental impacts, and ensure equitable access to our transportation systems and travel options to serve the needs of everyone.

Goals and Objectives

Predictable, reliable, and clean choices for the movement of people and goods underpin a strong economy and healthy, vibrant communities. Further, increasing equitable transportation mobility helps everyone meet their daily needs with dignity and independence.

Around the world, efforts to balance the benefits of new mobility modes while incorporating new technologies to bring more livable communities to everyone are a goal anyone should be able to get behind. By capturing different aspects of the transportation ecosystem—such as e-commerce and freight, passenger transportation, infrastructure, connectivity, data, and the workforce that supports it—a measure of adjacent but interrelated impacts can be realized, and the rethinking of how we best write policy, deploy these services, and govern can happen.

In that spirit, a small group set out to build a comprehensive new vision of our transportation future that is free of incumbent thinking and the same, tired and templated approaches to policymaking used for decades in this sector.

Approach and Challenges

To tackle this, the Commission on the Future of Mobility was created. Launched in October 2020 to reimagine how transportation systems are managed and optimized for better outcomes that benefit people, this coalition brings together select leaders from private industry, government, and academia. Its mission proposes creating a new vision of transportation policy for the movement of people and goods and the development of more inclusive communities across North America, Europe, and Asia.

Setting clear objectives to guide the policy research that would inform the advocacy framework, the group drew consensus on the importance of mitigating the global climate crisis and improving air quality by lowering emissions, building foundational structures that can enable technology to solve society's fundamental mobility problems. And last, but certainly not least, the research sought to foster people-centric communities and an enhanced quality of life for all.

Ahead of the formal launch, time was spent canvassing and recruiting to develop a board of leaders and advisors that would help drive the Commission's advocacy efforts. These leaders and this work would help people better understand why things need to change in the industry and would help shape the necessary policy recommendations to achieve that change.

Finding diverse, unique expertise and individuals with a personal passion across the transportation, energy, environmental, and technology sectors would be necessary, along with public sector leaders spanning three continents. A powerful global coalition was formed, coming together amid a transportation technology revolution this world has not seen since the invention of the gas engine automobile.

As of April 2021, the Commission on the Future of Mobility is proudly led by Commissioner Co-chairs:

- Ford Motor Company President and CEO James Farley, Jr.
- California Air Resources Former Board Chair Mary Nichols.
- Transdev Group Chairman and CEO Thierry Mallet.
- National Academies of Science, Engineering, and Medicine Board Chair on Energy and Environmental Systems and Carnegie Mellon University President Emeritus Jared Cohon.

Alongside the Commission's Co-chairs is an esteemed group of individuals spanning the transportation, environmental, government, and business ecosystems and helping to shape this critical project and set a new global perspective on the future of mobility.

Commissioners:

- Ola Cabs Co-founder and CEO Bhavish Aggarwal.
- Valeo CEO Jacques Aschenbroich.
- Former EU Commissioner for Industry and Entrepreneurship Elżbieta Bieńkowska.

- Former EU Commissioner for Energy and Climate Action, Miguel Arias Cañete.
- Bipartisan Policy Center President Jason Grumet.
- Allianz SE Member-Management Board Jacqueline Hunt.
- Wallenius Wilhelmsen Logistics ASA EVP and COO of Logistics Services Michael Hynekamp.
- Goodyear Tire and Rubber Company Chairman, CEO, and President Richard Krammer.
- Qualcomm Inc. CEO Steve Mollenkopf.
- Vision Ridge Managing Partner Reuben Munger.
- Hyundai Motor Company Global COO and Hyundai Motor North America and Hyundai Motor America President and CEO José Muñoz.
- Former Executive Director and Chairman of the Sierra Club Carl Pope.
- Oak Foundation Vice-Chair of the Board of Trustees Kristian Parker.
- Cox Automotive President Stephen Rowley.
- Arrival Ltd. CSO and President Avinash Rugoobur.
- FedEx Chairman and CEO Frederick W. Smith.

The Commission on the Future of Mobility set out to disrupt the traditional transportation policy playbook to put people first. By enacting bold strategies and breaking down silos, it seeks to propose solutions that address the growing autonomous, connected, electric, and shared mobility dimensions at our doorstep, combined with the need to foster livable communities that enhance economic opportunity and quality of life for everyone.

Conducting data-driven research across six essential areas of mobility and community infrastructure, the Commission hopes to bring the freedom of cleaner movement to all people and increased access to goods. This system can, in turn, be safer, more resilient, and more informed by technology than today's existing system, with greater access and equity for all users.

Drilling down, the focus areas of the Commission's policy research explores technology advancements in freight transportation, including how automation, drones, and other technologies can reduce the impacts of climate change. Right-sizing of vehicles and mode shift in passenger transportation sets out to address what it will take to incentivize people and businesses to use the best option for their trip while also exploring how vehicle regulations pin us into overdesigned vehicles for most of the trips that we take.

Furthermore, looking at the massive amounts of data being generated and determining how that data can empower governments to do their jobs better is paramount as society becomes increasingly connected. Data stewardship is a necessary investment if the public sector is to leverage this growing resource. And similarly, redefining how the public and private sectors can better work together to achieve outcomes in the best interest of people.

On that same trajectory, we must build our cities and communities to enable people and offer better utilization of our transportation assets. Smart infrastructure and a better street for all is imperative to support what has become the accelerated rise of e-commerce and a transformed and borderless digital economy.

Finally, as innovation flourishes, the effects on the sector's global workforce inevitably evolve. Jobs in the automotive and transportation sector will be eliminated, and as it scales, automation will have similar impacts on individual economic opportunity. The Commission will dive into what a just workforce transition needs to support the millions today who earn their livelihoods enabling the movement of people and goods and to help ensure they are not left behind. Pathways to reskill and upskill workers for the new mobility jobs and adjacent jobs created by this technological shift are essential to a robust global economy.

This policy research will shape the framework for the Commission's global advocacy efforts to affect legislative and regulatory policy change that accommodates a new era of mobility and modernized transportation infrastructure to serve the needs of everyone.

Equally important to defining the Commission's approach to mobility policy solutions fit for a more modern, global economy was the significance of hiring a team to help build and lead this new global project who were passionate about the future of mobility and were rising stars in their own fields.

The Commission is a project of SAFE, a nonpartisan, nonprofit advocating for energy and transportation policies that improve America's energy security and the responsible use of domestic energy resources. SAFE hired industry veteran Alisyn Malek to lead the Commission as its Executive Director, and she quickly set out to build her leadership team in both the policy research arena and strategic communications to support this type of new public campaign. In the first quarter of 2021, she brought Marla Westervelt and Ashley Simmons onboard. Together, this women-led executive team is building out a global staff to work alongside the Commission's leadership and fulfill its goals.

We are not the first generation to turn to new tools and new technologies to solve the problems plaguing our transportation systems, cities, and communities around the world. But are we clever enough to learn from past mistakes to do it right this time? The Commission on the Future of Mobility believes it has come to be at this pivotal moment of technological change and environmental necessity to provide a new vision and way of thinking about mobility policy that benefits everyone, everywhere.

One thing is certain: the future of mobility offers enormous promise. This project and its supporters are determined to ensure that the new transportation technologies and services this era creates help carry our communities forward to a future that is safe, smart, clean, and more resilient, but most important, puts people first. It's our future, let's get moving.

About the Team

Alisyn Malek is the Executive Director for the Commission on the Future of Mobility where she brings extensive business development, operations, and industry experience to inform her work. From EV product development and corporate venture and strategy at General Motors to early-stage startups as the co-founder and COO of May Mobility, where she led the building of an autonomous vehicle transportation solution that would solve urban transportation challenges ahead of the competition. Malek is also the founder and CEO of Middle Third, a boutique consultancy focused on mobility strategy. Malek was recognized as a top ten female innovator to watch by Smithsonian in 2018 and named a top automotive professional under 35 to watch by LinkedIn in 2015 for her work in cutting-edge product development and corporate venture.

Ashley Simmons joined the Commission on the Future of Mobility as its Director of Strategic Communications in March 2021. With more than a decade of experience specializing in public relations and public affairs, Simmons has worked with private sector-led advocacy organizations, startups, and nonprofits to develop communications strategies that elevate reputations, shape advocacy efforts, and expand global influence. She is also the founder and CEO of REVVIT Public Relations and has led communications teams internally at DC's most notable advocacy organizations representing the technology, telecommunications, and transportation industries. She has extensive experience developing media strategies, transforming brands, and carrying out integrated communications campaigns to educate the public and policymakers on tech-based legislative and regulatory policies affecting people and communities around the world.

Marla Westervelt is the Director of Policy for the Commission on the Future of Mobility. With nearly a decade worth of experience in the transportation sector, Marla spent her early career working at the Eno Center for Transportation, where she led research efforts from reforming air traffic control governance to rethinking the way we pay for infrastructure in the United States. From there she headed to LA Metro, where she was a founding member of the Office of Extraordinary Innovation, leading research efforts. She then left to be an early member of Bird Rides, leading global data-sharing policy and government-facing business development efforts. Most recently, Marla helped in the initial business development of MobilityData. This broad wealth of experiences uniquely positions Marla to drive a thoughtful and rigorous research agenda, exploring and seeking to define the future of mobility.

Remixing Innovation for Mobility Justice, 2020

Rachel Zack, Director of Policy, Remix
California

Reprinted with permission from © Remix.

Background

Across the United States, DOTs are rethinking their roles as road builders and working to integrate more equitable practices into their infrastructure planning and project delivery process. The "Remixing Innovation for Mobility Justice" project worked with the DOT leaders, policy experts, and mobility justice advocates to cocreate technology in an effort to accelerate demographic analysis and reduce time spent working on administrative tasks and planning in order to provide more spending and time dedicated to community engagement.

One of the policies in place at the federal level to make sure that community efforts and federal dollars are aligned, sufficient, and inclusive is Title VI of the Civil Rights Act of 1964. It states that "No person in the United States shall, on the ground of race, color, or national origin, be excluded from participation in, be denied the benefits of, or be subjected to discrimination under any program or activity receiving Federal financial assistance." All engagement efforts must follow these requirements in order to receive and maintain eligibility for federal funding. Though these efforts are often conducted and can be measured with some data, the work is often a combination of both qualitative and quantitative efforts, making it a large lift to measure and analyze. DOTs are able to budget for the engagement efforts for projects, but a Title VI analysis is often something nice to have, but largely viewed as a fringe cost.

Though not required as a part of the federal requirements for engagement, an analysis of Title VI efforts helps tell the story of how a service provider can identify gaps with technology and accessibility. Often, providers can recognize service gaps or trends from their riders, but they are not always able to identify the root cause or solutions. This is where an analysis comes in to help service providers understand how effective their efforts are. The analysis also allows municipalities to assess equity at different scales of infrastructure.

In the introduction for the final report of "Remixing Innovation for Mobility Justice," we discuss the institutionalization of whiteness from a federal level and note that the planning field has deep roots in discriminatory and exclusionary practices. This set the tone for the remainder of the report and how racism existed within federal policy. It also showed how the work done now and work planned to do tomorrow will be better when it is framed with an equity lens, not only serving the communities that use services such as public transit or shared mobility, but working to support them, and leave the community members better.

Goals and Objectives

Remix is an organization that works to determine the role of software in delivering equitable infrastructure, to design tools and features to meet the needs of practitioners and advocates, and to increase awareness around those who practitioners plan for, while adequately understanding their needs and finding ways to execute and implement solutions quickly.

The tools like Explore, Remix assists clients with Title VI compliance to:

- Measure inequities.
- Develop a framework to help motivate practitioners to implement equitable planning.
- Apply expertise and tools through case studies to conduct real-time data collection.

Approach and Challenges

Over the course of a year, Remix and TransForm worked together to better understand equity practices and collaboration in the public and private mobility sectors. The project team included equity advocates, DOT planners, and technologists who worked together with technology to develop software solutions to narrow the gap between equity practices and equity analysis. Remix developed Explore, giving planners faster access to transportation and infrastructure datasets that could be used for decision-making. TransForm started their portion of the research by conducting an equity survey for advocates on how they would be able to use Explore.

The headlines of 2020 affected the planned workflow for this project. The movement for Black lives and the pandemic both directly affected the process to complete the project partnership, and it was not possible to fully implement the case studies as designed. Through this year-long partnership, the Remix and TransForm teams worked together through a variety of outlets to determine the best ways to support their mission. During this time, Remix led a week-long design sprint with DOT leaders, policy experts, and mobility justice advocates to cocreate technology that would accelerate equity work related to infrastructure planning and development.

In working with equity advocates, Remix was able to validate the minimum viable product and to inform clients on which datasets would need to be included for a comprehensive view of a community's needs. After meeting with the advocacy group, the next step would be the application of the findings. Together, they co-authored a final report highlighting how, when, and where to use technologies in equitable infrastructure planning.

Below are summaries of the three case studies identified in the report that display how Remix can support equitable mobility solutions. These case studies were included in the final report as examples of how equitable planning can be conducted and what considerations should be taken.

Case Study 1: Making Sure a New Park Serves Low-Income Earners in the Surrounding Area

The first case study focused on a new park serving local low-income earners in the community. The new park would be built at 700 E. 98th Street, Los Angeles, California (Figure 8.1). The need for public and greenspaces was heightened during the COVID-19 pandemic as people struggled in new ways with physical wellness, mental health, and quality of life and with the new challenges associated with isolation. According to the Trust for Public Land, "100 million Americans do not have a park within a 10-minute walk" from their homes, leaving one in three Americans reliant on nonpedestrian forms of transportation to enjoy the benefits of being outside.

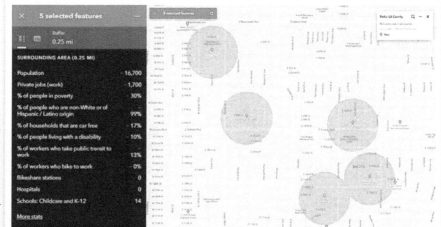

FIGURE 8.1 Five selected features for parks in LA County.

Case Study 2: Equitable Changes to Bus Service that Meet the Needs of Riders

Transit agencies have experienced significant budgetary challenges in the face of the COVID-19 pandemic as ridership and funding slowed and became less available, while costs to maintain service centers and vehicles increased. All over the country, transportation providers struggled to serve the needs of transit-dependent users, whether it was a ride to the doctor, to the grocery store, or to see family. Limited options were forced to become even less available as a result of the pandemic. As a result, agencies expect to see an increase in people owning cars and driving personal vehicles, both increasing their household transportation costs and leading to what Remix refers to as the "transit death spiral." The transit death spiral is a trend indicative in transportation networks in which reducing transit service results in lower ridership, which then leads to more service cuts, thereby compounding transit challenges.

TransForm and Remix worked together to see who was most impacted by the service cuts in Oakland and the Bay Area in California. They used the AC Transit Route to analyze service, although they do not currently have an active bus route. Remix datasets were used to interpret the distribution of low-income people along the route (Figure 8.2).

FIGURE 8.2 Potentially impacted routes and residents for service change.

Case Study 3: An Exploration to See if Oral Histories Match Data

In Fresno, California, Thrivance Group worked with the community and government partners for displacement-avoidance policies to avoid the displacement of very low and low-income households by developing and leading the "Displacement Avoidance Plan Design," an implementation for the "Transform Fresno Plan." The team is working to ground oral interviews and histories with data in order to show how some accessibility functions in infrastructure investment might actually be at the root of displacement to certain populations. Although most infrastructure projects have data-driven approaches to prove community benefit, planners rarely return to the project postcompletion to analyze the impacts and evaluate whether the proposed project components and benefits were realized. The approach for measurement was to use Remix to compare the ACS data (U.S. Census Bureau) in order to explore if it supports the qualitative data (oral histories) as a new method for longitudinal equity impact assessments.

Results

Reprinted with permission from © Remix.

TABLE 8.1 Using Remix Explore to further understand community needs.

	Case study 1	Case study 2	Case study 3
	Making sure a new park served low-income earners in the surrounding area	Equitable changes to bus service that meet the needs of riders	An exploration to see if oral histories match data
What Remix Explore can do. Quickly answer questions that would otherwise take an analyst hours or days.	Make the connection to environmental racism's cumulative and disproportionate impacts on BIPOC people. Ten minutes into her case study, Tamika was able to answer her initial questions and discover new ones in real time and answer them just as quickly.	Quantify the number of transit-dependent riders living within walking distance to a specific bus line being targeted for service reductions. Quickly investigate potential impacts for the agency reducing service.	Create data visualization on a map with demographics statistics. Quickly enable the change between ACS data from one year to another.
What Remix Explore cannot do. Make a decision or determination for an issue or concern.	Talk to community members. Too often community feedback and opinion are not valued in the same way that empirical or quantitative research are valued.	Put together different types of analysis to come up with an advocacy strategy. Make sure transit planners and agency decision-makers are asking the right questions when they plan to service changes.	Illustrate direct benefits or risks to populations of concern, through some functionality can show trends over time. Explore does not capture oral history as a data point to help build a case for setting spoken narrative besides qualitative data.
How Remix Explore works best. Helps advocates influence decision-making processes.	When going to decision-makers who have discretion and resources to build new greenspaces, advocates could be equipped with the data and ability to show the impacts of different decisions.	Equip advocates to understand and communicate the trade-offs that transit-oriented planners make when deciding on service changes. Often these impacts on riders are not present to governing bodies when they are making decisions on service cuts.	Help operationalize a planner's workflow. Thrivance needed specific data points to understand displacement impacts unique to their project area. Explore allows them to see the population represented by statistics to then gauge implications cross-referenced to policy or other factors.

Reprinted with permission from © Remix.

Case Study 1

After a base analysis was conducted, the team concluded that the proposed park would be located in a community with a disproportionately high population of people of color. The Remix Stats Grab tool was applied to an LA City Parks District Layer and used to compare this future public park to the 39 other parks in the district. This analysis concluded that this area was populated with 99% people of color, car-free households, and transit mode share users were higher than the city at large. This tool also identified that there were no hospitals existing in the corridor. This tool showed that the proposed park will serve close to 11,000 people who currently do not have access to a park within a half-mile radius and 13,7000 people overall.

In addition to offering the solution and findings on the case study, the report spells out how to achieve a similar finding without Remix. Below is an excerpt from the report:

> The best way to determine if this is a good site for a park and if residents would benefit from it would be to talk to community members. However, too often community feedback and opinion are not valued in the same way that empirical or quantitative research are valued. As a result, when trying to determine if a particular site is good for a park, planners often use research from academic institutions or from prior government planning processes for green space in the area. This could include planning documents from a local council office or from the Department of Parks and Recreation. There is also quite a bit of discretion in where parks can be built that can be determined by elected officials and developer.

What makes Remix and the Explore tool different from other platforms aimed at providing data to drive planning decisions is that the team recognizes that a tool like this could create another layer of barriers that could further complicate the congruence of equity being applied across public planning projects. This is where Remix pulls in their collective experiences with policy and calls on government agencies to share their findings, training, and tools with community-based organizations. This would also support relationships between the public and the private sector.

Case Study 2

By tapping into the tremendous amount of data made available by Remix, the team was able to identify that about 22% of the 7,626 transit riders along this line were at or below the poverty level. Remix took the data and compared it against different indicators of poverty (100% of the general poverty level, 200% of the federal poverty level, etc.) to account for the high cost of living in the Bay Area. By zeroing in on certain geographies, they were able to find indicators that showed a specific neighborhood along the route with 30% limited-English speakers that was 84% nonwhite, which was higher than the sample bus route.

"Jane" is an isochrone feature inside Remix Explore, demonstrating that the residents in the study community, dependent on this line, have access to most of downtown in 45 minutes to 1 hour. An isochrone map in geography and urban planning is a map that depicts the area accessible from a point within a certain time threshold. An isochrone is defined as "a line drawn on a map connecting points at which something occurs or arrives at the same time." Through Jane, the team was able to analyze how the communities around this bus line would be impacted by potential service adjustments. Because service is centered on the dominant 9-to-5 commute, many low-income workers like hospital and service workers who are dependent on reverse-commute patterns are most at risk of losing their transportation access when services are reduced.

Case Study 3

Thrivance provided a shapefile (a geospatial vector data format for GIS software) of the infrastructure projects that were to be compared. This file was uploaded into Explore for analysis. Remix displayed how the infrastructure projects that were meant to improve community mobility options had not changed mode split—the percentage of travelers using a particular type of transportation or number of trips using said type in the local area. These results showed that the lived experience of the community did not match the expected outcomes from the infrastructure project planners. This process proves that technology can be applied in new ways to augment existing datasets to tell a more comprehensive and actual story. In this case, the experiences of black, indigenous, and other people of color (BIPOC) were not originally represented in the data, so the combination of qualitative and quantitative data helped display the actual results of the infrastructure "investment" to the community.

The results for this project have been tremendous. Agencies that wish to roll out more demographic analysis have turned to our tools to do so as the time burden has shifted from days to minutes. As proven in the above case studies, Remix helps communities, agencies, and organizations achieve the following outcomes:

1. Data definition for community input, guidance, and/or leadership.
2. Flexibility and adaptation to community needs as they change and grow.
3. Providing the appropriate tools and resources to those people assigned to conduct community engagement.
4. People-first equity planning rather than infrastructure planning.
5. Expedited analysis periods by using tools that efficiently and quickly identify results.
6. Establish an authentic community engagement strategy that reconnects the decision-maker and the community.

About the Team

Rachel Zack is the Director of Policy at Remix, a transportation planning software startup under Via. Her passion for empowering planners with access to insights for better decision-making stem from her many years of advising on cutting-edge transportation problems such as TNC regulation (Uber and Lyft), road-usage pricing, congestion pricing, express lanes, and planning for AVs.

Before joining Remix, Rachel launched the Innovative Mobility arm of WSP, one of the largest global professional services firms, where she led alternative transportation programs around the Bay Area and advised city and state agencies on strategic planning for new mobility. Her work in the field received the Caltrans Excellence in Transportation Award and WTS Innovative Transportation Solutions Award. Rachel has a master's degree in City and Regional Planning from the University of Pennsylvania and a B.A. in Growth and Structure of Cities from Bryn Mawr College. As of June 2021, Zack now works as the Head of Partnerships at Felt.

Additionally, this project was developed and delivered, by and large, by a diverse group of strong and devoted womxn: Tamika Butler Consulting, Thrivance Group, Greenlining Institute, Catholic Charities, and Elemental Excelerator, who all believe in the role that transportation can play in increasing access to opportunity. This work could not have been completed without the hard work of Jamario Jackson and Darnell Grisby of Transform and Jonathan Pruitt of the Catholic Charities of the Diocese of Stockton, California. This devotion and urgency does not stop after the tools are built or the paper is written. It is up to all of us to continue to push for the advice to be acted upon and the tools to be incorporated into the workflow.

9

Sustainability and Climate Resilience

When we think of a national infrastructure resilience policy, the first projects that jump to our minds might be a sea wall or road elevation related to flood mitigation. The process of keeping our residents safe during heat and cold extremes has fewer dollar signs associated with it. But especially because extreme heat is the biggest climate change killer, ensuring that our public transit system—especially buses—remain funded, and designing them to allow them to flex into cooling or heating stations not by accident but by specific solution should be a key part of resilience funding.

—Joyce Coffee, President, Climate Resilience Consulting

Did you know that it takes an estimated 39,090 gallons of water to make a car? This is an element inherent in the existence of the car but nearly invisible to the purchaser. The lack of footprint visibility probably will not directly impact the average consumers' purchasing decisions, but that footprint is very much there for every item produced and consumed. This consumption of natural resources and man-made materials is exploited in every single decision we make and, as a result, aggressively and significantly contributes to the rapid impacts of climate change.

We want to make something perfectly clear: climate change is not religion, it is science. We have to stop "believing" in climate change and start managing it. It will not be easy, and it will not be comfortable, but it is necessary in order to move humanity forward.

Whether it is the car we choose to drive, the kind of home we choose to live in, or the dinner on our table, there are "hidden" costs in each: their environmental impacts. There are so many small decisions that we make on a daily basis that have a much higher cost than meets the eye, often at the expense of convenience, but it is those exact decisions, made individually and collectively, that bring me to this point: *if we do not collectively work together to combat the climate crisis, the rest of the incredible progress in this book will not matter.* The only thing more expensive than the cost of sustainable investment or the increased costs associated with a more sustainable lifestyle is the cost of doing nothing.

Unless we reverse the warming of Earth's atmosphere, significantly adjust consumer behavior, adapt global markets to clean energy and decarbonization, and work together, we will continue to be in a global war against Earth, herself. We are already seeing the traumatic signs, as she lashes out through climate events, such as hurricanes, heat events, fires, coastal flooding, sea-level rise, earthquakes, mudslides, and tornadoes, and the public health and economic implications that follow.

The work toward a more sustainable future will not be successful unless it involves every living thing. There is a critical equity piece to this work, and without it, climate action will not be a climate solution. The privilege instilled in those of us fortunate to make the choice to abstain from eating meat, choose to drive an EV, or even upgrade our incandescent light bulbs to LED is not an option for everyone. Truly, these changes might not be enough to create the drastic change needed to reverse the warming of our planet. There are many large problems that need to be addressed through policy and legislation solutions, like the Paris Accord treaty, with funding sources like a carbon tax and with massive adoption of our existing systems to support more options favoring multimodal transportation, sustainable community development, and aggressive conservation efforts. The UN suggests these building blocks provide a realistic approach to respond to climate change mitigation: adaptation, technology transfer, and financing. These "building blocks" will work together to reduce, store, remove, and offset GHG emissions.

GHG emissions come from a multitude of sources, but there is one key producer: transportation. Transportation alone is responsible for 25% of the world's energy-related emissions of GHGs. A GHG is released into the atmosphere as a result of burning fossil fuels such as gasoline and diesel. This buildup of hydrofluorocarbons, nitrous oxide, methane, and CO_2 warm the atmosphere of Earth, causing the extreme weather events we are seeing today. Every year for the past seven years has been the warmest year on record, a trend that continues to grow and show signs of a major atmospheric disruption. Climate change is already happening, and we are seeing it in the chaotic weather events that disrupt our lives, though we will all experience it differently—and some more than others.

Women and BIPOC disproportionately experience the negative effects of climate change. Across the globe, women are more vulnerable than men due to restricted access to economic participation, education, and decision-making opportunities. Additionally, gender roles and pay gaps continue to force women into unpaid caretaking roles, often as sole caregivers for others. In these roles, women have significant exposure to the climate issues associated with food, water, and health-care access. Additionally, both women and BIPOC continue to be underrepresented in the professional fields associated with resilience, sustainability, and climate science. Though enormous efforts are being made to advance spaces for women and girls in STEM, it feels as if these efforts might be too late—although we hope they are just in time.

In a study completed by McKinsey,[1] a "full potential" scenario is mentioned in which women play the same role in the labor market as their male counterparts. In this scenario, it is expected that the global annual gross domestic product (GDP)[2] would grow by 26% by 2025. GDP is a monetary measure of the market value of all the final goods and services produced in a specific time period. This 26% represents $28 trillion. Though much more investment is required to manage and reverse climate change, the current global spend sits at $585 billion annually as of 2020. Increasing women's share of the market in the workforce is enough to finance climate change mitigation, adaptation, and technology transfer multiple times. If this can be realized at equal gender representation of GDP, imagine how much could be solved with more women and more gender diversity present in leadership roles of the sciences.

In this chapter, we will learn from some of the women who have dedicated their careers to this fight and have been working to protect our Home for decades. From the lush coasts of Hawaii to the post-Harvey recovering City of Houston, we will see how large-scale change from the transportation sector can protect the homes, lives, and communities of Americans; we also will see how tragically destructive the absence of those changes can be. Additionally, we will see the solutions being deployed in U.S. cities as we continue to build to meet the demands of our busy world. As we adjust to the world after the screeching halt from COVID-19, we intend to build back better by prioritizing clean energy and bringing more diverse voices to this critical conversation. As gender diversity grows, and as women and underrepresented groups of people become louder and more present in the science community, we might just turn this thing around—because we have to.

Achieving Sustainability and Equity through Electrified Micromobility

Melinda Hanson, Co-founder and Principal at Electric Avenue and former Head of Sustainability, Bird
New York City

---

[1] https://www.mckinsey.com/~/media/McKinsey/Business%20Functions/Sustainability/Our%20Insights/McKinsey%20on%20Climate%20Change/McKinsey-on-Climate%20Change-Report.pdf
[2] U.S. Bureau of Economic Analysis (BEA), "Gross Domestic Product," Retrieved February 23, 2019, www.bea.gov.

Background

In 2020, Electric Avenue launched in New York City to help mobility providers scale their fleets, identify policy gaps, and advance opportunities to align with and serve more users. Roughly 60% of car trips in U.S. cities are less than 5 miles, and 20% are 1 mile or less. Bicycle ridership has been slowly and steadily growing in New York City, but it has not kept pace with growth in car trips. While the majority of distances driven by car are short, these trips are often made too inconvenient or time consuming by walking or using public transit. Electric Avenue works with private companies, NGOs (nonprofit organizations that operate independently of any government), and the government to help bridge the gap between 1- and 5-mile vehicle trips by encouraging first-mile and last-mile connection solutions through electrified micromobility.

The two large issues that Electric Avenue aims to address are equity and climate change. In most U.S. cities, having a car is a requirement for getting around, though this is not an option for everyone due to accessibility, economic opportunity, infrastructure, or preference. Car ownership costs the average American approximately $9,000 annually, and with the average American income sitting at approximately $31,000 per year, owning a car can cost an individual nearly 29% of their income.[3] In some locations, and for some individuals, the cost can increase to as much as 50% of their annual income. Overregulation of micromobility and smaller modes of transportation that can help offset those costs is an equity concern and leads to transportation injustice. Concurrently, these larger, more expensive modes of transportation—like personally owned vehicles—emit far more in GHG emissions than electrified scooters or e-bikes. If only 15% of car trips were replaced with a more sustainable option, such as an e-bike, the total carbon emissions would drop by as much as 12%.[4]

Electric Avenue sees their role as a liaison between private and public sectors, working to get sustainable mobility solutions deployed quickly and equitably. Public-private mobility partnerships face a lot of friction due to the nature of partnerships and funding structures, but shared electric micromobility shows great promise for fighting congestion, improving transportation access, and fighting climate change—though the pace of change is inadequate. Electric Avenue was launched to help bring sustainable transportation to more markets faster and more efficiently. Transportation companies consistently underinvest in policy advocacy and public affairs. Electric Avenue helps with this.

Goals and Objectives

The main goal for Electric Avenue is to serve as a conduit between public and private entities, with a mission to achieve a 20% mode share (the percentage of travelers using a specific mode) in cities with two- to three-wheel EVs. Across the United States, there is about a 1% two-wheeled mode share across the nation, with the highest being

---

[3] https://www.bts.gov/content/average-cost-owning-and-operating-automobilea-assuming-15000-vehicle-miles-year
[4] https://www.bloomberg.com/news/articles/2021-03-31/switching-from-cars-to-bikes-cuts-commuting-emissions-by-67

about 6% in Portland, Oregon.[5] By serving as experts in the space and being able to engage in calm conversations in the mobility field, they are able to help negotiate and advance better regulations to streamline public-private partnerships and encourage innovation and creativity to meet and exceed the transportation demands of municipalities and their communities.

Electric Avenue seeks to achieve the five main objectives listed below:

1. Expand vehicle types to more groups of users: A one-size-fits-all approach does not work when it comes to transportation and mobility. To meet the needs of different user types, this organization works with clients to better understand how to design and deploy vehicle types that are accessible to a broader swath of the population. These vehicle types include sit-down tricycles as well as electric devices that attach to existing mobility vehicles of users, such as wheelchairs. Electric Avenue is currently working with shared micromobility company Voi to explore the possibility of deploying Klaxon, a standalone electric device that is attached to any wheelchair with a small hitch to adapt the vehicle and electrify it.

2. Expand connected and protected infrastructure for micromobility and pedestrian activity: A key theme seen time and time again in the mobility conversation is the deterrents that can make mobility solutions in practice unreliable and at times even dangerous. This can include bike lanes with cars parked in them, nondedicated transit lanes interrupted by ride-share staging, and sidewalks with, well, anything that is not a clear path. Electric Avenue works with clients to shape and promote policies to protect vulnerable street users, promote inclusive design in infrastructure, and develop new approaches to ensure safe and orderly deployment of new mobility devices.

3. Promote incentives and other smart policies to help micromobility to scale. Micromobility offers a great solution for reducing carbon emissions, yet it has not enjoyed any of the subsidies, incentives, or public infrastructure investments to support its use in the way EVs have been supported. Electric Avenue works independently and through its clients to advance conversations at the city, state, and federal levels to bring parity to the field.

4. Promote financial accessibility across transportation options: Affordability of a variety of transportation options is critical to the success of micromobility programs. New York City, like many other urban areas, is challenged with serving populations that depend on cash for spending and are credit or bank insecure. Some cities propose subsidized monthly passes for users with income limitations who already receive benefits such as public housing; others are exploring following a more European option of adding transportation costs to utilities. Electric Avenue is also working to pilot transit connectivity solutions so that users can use a transit pass to streamline payment.

5. Improve access for non-English speakers: Electric Avenue works with clients to make sure that any respective apps working to improve the other listed objectives are available to users in multiple languages. Popular languages that are initially recommended include Spanish and Mandarin Chinese.

---

[5] https://www.portlandoregon.gov/transportation/article/709719

Approach and Challenges

In 2017, scooters were dropped off in many urban centers, appearing literally overnight. Cities responded with strong regulations to limit or prohibit their use. Further, these new vehicle types existed in a "gray area" of the vehicle code, meaning it was unclear whether or not they required a license to operate or should be treated more like bikes. Conversations initially were focused on cleaning up codes to make scooters street legal and setting operational and safety parameters to ensure shared programs worked well for both operators and cities.

These surprise micromobility deployments were happening in many major cities across the country, and Electric Avenue was founded in 2020 after some of these trends continued to replicate. The founders recognized the potential to close the gap between transportation solutions and their respective policies.

In 2021, Electric Avenue is exploring how to help shift policy and behavior to reconsider how we move and whether that is through electrified transit, a car, scooter, tricycle, bike, or something better. Electric Avenue is uniquely positioned to achieve this goal by approaching these challenges with a team of experienced industry leaders who use a combination of private and public sector angles.

Courtesy of Melinda Hanson.

Results

In the first year following the launch of Electric Avenue, the organization saw a 90% success rate in helping clients win operating permits in the United States and Europe. The team attributed this success rate of adoption to the fact that they encourage all project work and clients to take a cities-first approach. Electric Avenue has been involved in elevating the profile of micromobility and building the advocacy community around micromobility to encourage its members to be as vocal as the EV world is.

One of the most important results Electric Avenue expects to achieve is community education. They participate in leading the global conversation around micromobility and transportation, lending their perspective of direct experience with transit operators. An example of one of the largest presentations by Electric Avenue was a workshop for the National Association of City Transportation Officials (NACTO) to support staff in cities to be equipped with the tools they need to write adequate requests for proposals for these projects. Electric Avenue help measure the impact of their services around equity, safety, and sustainability by collecting and providing data to other cities. Sharing best practices helps develop use cases (specific situations in which a product or service could potentially be used) and supports expansions of programs and infrastructure.

About the Team

Melinda Hanson is Co-founder and Principal at Electric Avenue, a public affairs and mobility strategy firm working with companies, nonprofits, and city agencies to move car trips to electric two- and three-wheeled options. She has spent her career working at the intersection of transport and climate, including as Head of Sustainability at Bird, Deputy Director at NACTO, and Portfolio Manager at the ClimateWorks Foundation. Highlights of Hanson's career include implementing tactical urbanism projects in Addis Ababa, Ethiopia; contributing to the master plan for a Bus Rapid Transit Project in Peshawar, Pakistan; and supporting research and advocacy to pass the Obama administration's landmark fuel economy standards. Hanson worked closely with her co-founder Louis Pappas to make this work possible.

Shaping City-Scale Urban Sustainability and Quality of Life with LEED for Cities

Hilari Varnadore, Vice President, LEED for Cities, United States Green Building Council (USGBC)

Background

Fifteen years ago, the USGBC partnered with ICLEI-USA and the Center for American Progress to develop a "green city index"; their commitment to action was supported by the Clinton Global Initiative and launched at Greenbuild in Chicago that year. Later the National League of Cities and National Association of Counties joined the collaboration to develop the nation's first framework and certification program for local government sustainability. Built by and for local governments, the STAR Community Rating System (STAR) was released in 2012, and the City of Tacoma, Washington, was the first to achieve certification in 2013.

A new nonprofit, STAR Communities was established to administer the rating system; the philanthropic community, namely, the Summit, Surdna, and Kresge Foundations, provided startup funding and operating support. More than 75 local governments across the United States certified in STAR between 2013 and 2017. The efforts of those cities and counties refined our common framework for sustainability and established the value of setting national standards for the field of practice.

Around that same time, USGBC released the LEED for Cities pilot, which engaged projects in the first-ever peer-to-peer benchmarking initiative. Powered by Arc, the pilot required that projects report data across 14 sustainability indicators. A performance score provided cities an opportunity to see how they compare to other places around the world across five categories: energy, water, waste, transportation, and human experience.

As noted, the USGBC was a founding partner of STAR, and it became obvious that the integration of the STAR and LEED for Cities programs could essentially lift up the best attributes of each program; together they could be an even stronger, more relevant framework and product. Where STAR captured key methodologies—like analyzing compact and complete centers and using maps to illustrate access and proximity to community assets and standards for green space, GHG emissions reductions, and income inequality—the LEED for Cities pilot added value with dynamic peer-to-peer comparisons. The two programs merged and LEED version 4.1 (v4.1) for Cities and Communities was released to the market in early 2019.

Goals and Objectives

Ultimately, the rating system and certification program aims to:

- Inspire leadership, transformation, and innovation in urban sustainability.
- Protect biodiversity and regenerate ecosystem services.
- Meet and exceed net-zero carbon, energy, waste, and water.
- Achieve livability, choice, and access for all, where people live, work, and play.
- Raise the standard of living and quality of life for humans all around the globe.

LEED for Cities helps local leaders to:

- Benchmark performance against national and global standards.
- Demonstrate leadership and commitment to sustainability.
- Improve the quality of life of their communities.
- Develop a culture of data-driven decision-making and performance management.

Approach and Challenges

A Vision for Cities: LEED v4.1 for Cities and Communities

LEED v4.1 for Cities and Communities provides a clear, data-driven approach to assessing conditions and evaluating progress, with a structure and point system that mirrors that of other LEED rating systems. The new, expanded system has eight categories:

- Ecology and Natural Systems.
- Transportation and Land Use.

- Energy and GHG Emissions.
- Water Efficiency.
- Materials and Resources.
- Quality of Life.
- Innovation.
- Regional Priorities.

Version 4.1 includes a traditional rating system of prerequisites and credits. Most prerequisites are performance based and are reported on the Arc platform. Others include base conditions such as requiring 100% of residents to have access to drinking water, solid waste management services, and electricity. The base conditions provide mapping and assessments of ecological and demographic assets for project context.

The credits in the rating system are composed of best practices and leadership standards that have been proven to move the needle on sustainability. Most credits include a mix of outcomes (quantitative) and strategies (qualitative) that the local government can report on. The certification program provides third-party verification of all projects using LEED Online and the Arc reporting platform for data collection and review.

USGBC's vision is that buildings and communities will regenerate and sustain the health and vitality of all life within a generation. LEED has evolved beyond buildings to the city and community scale to meet this need. In order to realize a sustainable future for all, the next generation of green building must focus on the development of smart cities and resilient communities. According to the USGBC,

> a city is comprised of a government (in some form), people, industry, infrastructure, education and social services. A smart city thoughtfully and sustainably pursues development with all of these components in mind with the additional foresight of the future needs of the city. This approach allows cities to provide for its citizens through services and infrastructure that address both the current needs of the population as well as for projected growth.

Our cities must champion equitable, safe, and healthy development policies, implement interoperable platforms and advanced technologies that improve the performance of their communities and cities, and continue to incorporate concepts like social equity and public health into city planning, development, and management.

LEED for Cities is a response to the need for a flexible, credible, and globally consistent way to communicate continuous urban sustainability performance across an array of objectives and to different types of stakeholders. The program harnesses the power of data to compare and benchmark aspects of performance across cities and communities and to roll data and metrics from the project level up to the city level.

The Impact of 2020 on Local Governments Engaged in LEED for Cities

In 2020, local governments grappled with the immediate and devastating effects of COVID-19 while confronting head on the vivid legacy of historic and systemic racism.

LEED for Cities played a constructive role in helping local governments navigate these challenging times.

Certification provided a gap analysis and roadmap to continuous improvement. The demographic and ecosystem assessment prerequisites revealed the distribution of community assets, highlighting where wealth and investment have been concentrated. Through this process, cities like Cincinnati, Ohio, and Louisville, Kentucky, made the correlation between low-income, formerly redlined neighborhoods and increased urban heat. These types of insights connecting environmental and health impacts led places like Charlotte and Fayetteville, North Carolina, and Cleveland, Ohio, to declare racism a public health emergency.

Sustainability practitioners and thought leaders came together to share their stories so that others could learn from their experiences. When residents were on lockdown and isolated in their homes, staff in LEED-certified Denver, Colorado, and Georgia, and Michigan spoke about ensuring equitable access to green spaces. When office workers stopped commuting and public transit ridership slowed, transportation and mobility leaders from Seattle, Washington, and LA, California joined thought leaders from the New Urban Mobility Alliance, the WRI, and the Natural Resources Defense Council, Inc. to talk about the impacts of those disruptions and how to move forward.

At the USGBC, we have the benefit of working with hundreds of local governments around the country and the world who have committed to data-driven decision-making. Cities are working within existing ordinances, policies, frameworks, plans, and decisions, which may have been made decades or generations ago. Committing to LEED demonstrates leadership and accountability to taxpayers, bond issuers, and other stakeholders. It requires an all-hands-on-deck approach and engages leaders across departments and throughout the community to work together to effect change. Change in community conditions often happens slowly at the city scale, so making adjustments now is critical to realizing long-term improvement.

Amid a global pandemic, more than 40 cities and counties committed substantial time and staff capacity to LEED for Cities certification. With the support of the Bank of America, the USGBC led the Local Government Leadership Program, engaging a national cohort of cities, towns, and counties in LEED certification together. Both as a cohort and as part of a national network, these local governments are demonstrating leadership and advancing ambition on sustainability, social equity, and resilience. They are taking important actions that are leading to reductions in GHG emissions, VMT, water consumption, and waste generation and are improving quality of life for all as a result of our engagement. Cities are taking stock and being more mindful of the impact of their decisions and investments through LEED for Cities.

Results

We are in the decade of action; the time to act is now if we expect to meet our 2030 climate goals. While urban sustainability requires us to commit to long-term planning with a focus on our impact to future generations, we also need to roll up our sleeves and get to work today.

LEED for Cities serves as the global standard for urban sustainability and, ultimately, will reach and empower global cities to become more sustainable, resilient, healthy, and just. In doing so, the program provides the ultimate vertical evolution for LEED ensuring impact beyond individual buildings and communities. The time to act on climate and social equity is now. Our goal is a sustainable world for all. LEED for Cities stands ready to listen, evolve, adjust, and improve to meet (and reflect) the needs of local governments worldwide.

As of March 2021, 120 cities and communities are LEED certified, and more than 210 projects around the globe are registered. Local governments certified to LEED v4.1 include Cincinnati, Ohio; Orlando, Florida; Royal Oak, Michigan; Santa Fe, New Mexico; Greensboro, North Carolina; Las Vegas, Nevada; Albuquerque, New Mexico; Costa Mesa, California; Rancho Cucamonga, California; Palm Beach County, Florida; Tampa, Florida; and Bloomington, Indiana.

About the Team

Hilari Varnadore brings more than two decades of experience in public administration and nonprofit leadership to her role as Vice President of LEED for Cities. Previously, Varnadore led STAR Communities as its executive/founding director. In that capacity, she deployed the first framework and certification program for local sustainability in the United States, the STAR Community Rating System. She has served as a chief sustainability officer and principal planner in local government and has led two nonprofit organizations as CEO.

Varnadore is invested in helping cities, towns, and counties use data to drive decision-making, investments, and community improvements that are more equitable and resilient. Her approach to sustainability is holistic and rooted in the triple bottom line (TBL).[6] The triple bottom line (or otherwise noted as TBL or 3BL) is an accounting framework with three parts: social, environmental (or ecological), and financial. Some organizations have adopted the TBL framework to evaluate their performance in a broader perspective to create greater business value. Her areas of expertise include strategic planning, facilitation, program development and administration, policy development, governance, stakeholder engagement, fundraising, marketing, and communications.

She holds an M.A. in Geography with an emphasis in Community Planning from Northern Arizona University and earned a B.S. degree in Environmental Policy from Unity College of Maine.

Varnadore serves on Metropolitan Washington COG's Climate, Energy and Environment Policy Committee, on the board of the Emerald Cities Collaborative, and on the EcoDistrict Advisory Council. She is a 13th-generation Marylander and a Girl Scout troop leader.

Additional Commentary

The development and deployment of LEED for Cities and Communities benefits from a global team of leaders, many of whom are also women. Hilari Varnadore, Vice

---

[6] Slaper, T.F. and Hall, T.J., "The Triple Bottom Line: What Is It and How Does It Work?" *Indiana Business Review* 86, no. 1 (2011): 4-8.

President for Cities at the USGBC, would like to recognize these women who are advancing sustainability and quality of life for all around the world as part of the LEED for Cities and Communities leadership team:

- Melissa Baker, Senior Vice President, Technical Core, USGBC.
- Mili Majumdar, Managing Director, GBCI India and SVP, USGBC.
- Reshma Kulkarni, Director, Technical Development, GBCI.
- Rhiannon Jacobsen, Managing Director, U,S, Market Transformation and Development, USGBC.

A Sustainable and Autonomous Future in Hawaii

Kathleen Rooney, Director of Transportation Policy and Programs, Ulupono Initiative

Kelley Coyner, Founder and CEO of Mobility e3 (Me3)
Hawaii

Background

In 2019, the UN made the statement that "We have just 12 years to make massive and unprecedented changes to global energy infrastructure to limit global warming to moderate levels," effectively providing a 12-year timeline to protect life as we know it (Figure 9.1).[7] For some communities, the impacts of climate change are already present—notably coastal states such as Hawaii. With 750 miles of coastline, Hawaii is one of the most vulnerable geographical areas already experiencing rising temperatures, volcanic warming, extreme flooding, tropical storms, sea-level rise, and ocean acidification. To protect both the environment and economy, the State of Hawaii prioritized sustainable policymaking and procedures and aims to have carbon positivity by 2045.

Tourism (pre-COVID) accounted for $17.7 billion in revenue in 2019, according to the Hawaii Tourism Authority,[8] making a strong transportation network critical to meeting the needs of moving people but also to the economic health of the state.

The threats of climate change combined with Hawaii's commitment to providing a safe and healthy environment that islanders and visitors can enjoy has led to an enterprising response pushing the state toward economic and environmental resiliency through adaptive action of their mobility network.

Transportation is recognized as a key contributor of GHG emissions, making it a first stop for climate mitigation. Hawaii-focused impact investing firm Ulupono Initiative, in partnership with Me3, worked to outline strategies that would prepare

---

[7] https://www.vox.com/2018/10/8/17948832/climate-change-global-warming-un-ipcc-report
[8] https://www.hawaiitourismauthority.org/news/news-releases/2020/hawai-i-visitor-statistics-released-for-2019/

FIGURE 9.1 A public electric vehicle charging station on the Island of Hawaii.

Hawaii to lead in AV and mobility technology to support a healthier and environmentally resilient state.

Goals and Objectives

Ulupono Initiative focuses on five key strategies to stretch environmental responsibility across the islands through areas including legal and safety, policy, infrastructure investment, planning, and mobility innovation. To achieve this, the team advocated for A2CES—an approach that prioritizes an "Accessible, Automated, Connected, Electric, and Shared" mobility future that promotes equity, the environment, and economic opportunity for all in Hawaii.

The roadmap to this future includes confirming the legal framework for safe operation of A2CES vehicles on roads. The legal guidance of these vehicles includes safety regulations and requirements. Powering the system is another key component of the framework. For complete A2CES readiness, the grid and power sources need to be prepared to support the network of vehicles and do so, powered by renewable energy. Further, policies should support public involvement in the planning process to ensure equitable and environmentally just transportation networks that promote economic vitality.

Since A2CES is a multimodal approach, all supporting infrastructure should be updated to accommodate the technologies of the system. This readiness should be baked into public and private sectors investment including rail stations, rail catchment areas, station access, and roadway and multimodal pathways. EVs, including AVs, should be considered in all aspects of planning so that the vision of A2CES is recognized and realized through land use, economic development, and transportation planning.

Approach and Challenges

Beginning in fall 2018, Ulupono Initiative collaborated with local and state government agencies to bring stakeholders together to lay the groundwork for accelerating the development and deployment of AVs to maximize associated benefits to the people of the state. Me3 helped to articulate an A2CES vision of mobility and recommend a framework of private- and public-sector collaboration. This framework provides an actionable roadmap to achieve economic sufficiency for Hawaii, reach renewable energy goals, and give those who cannot drive more autonomy through mobility.

The findings were outlined in a report titled "Framework for Hawaii's AV Future: Accessible, Automated, Connected, Electric, and Shared" that was published in May 2020.

During research and development (R&D) of the report, the planners acknowledge that Hawaii is in some ways already a leader in autonomous transportation. For example, Governor David Ige's signing of Executive Order 17-07 (Autonomous Vehicle Testing) made Hawaii the first state to identify its airport for the deployment of autonomous shuttles. In addition, Honolulu is moving closer to opening the first stage of its automated rail transit system, joining Vancouver, British Columbia, and Copenhagen, Denmark, with an automated elevated metro that will be the first fully driverless system in the United States. However, the planners cite disruptive technologies such as AVs for having turned upside down the regulatory paradigm—meaning long-held planning models that prioritize single-occupant, gas vehicles are no longer adequate—and the effect of Hawaii's accomplishments is lessened due to a lack of clarity over the conditions for deploying these vehicles on public roadways.

Results

Since being established in 2009, Ulupono Initiative has funded and supported organizations and programs across various sectors including renewable energy, local food production, freshwater management, and waste management. The majority of Ulupono investments have focused within the sectors of food and energy (including clean transportation), 47.5% and 48.5% in 2019, respectively. It is also worth noting that, with Hawaii importing approximately one-third more gallons of fossil fuels for ground transportation as for electricity production, Ulupono has begun focusing on clean transportation as a sector in its own right, committing additional resources accordingly.

This investment of resources supports the A2CES framework to create a mobility system that promotes Honolulu's autonomous rail transit system, a backbone for a larger system that aims to connect people to economic opportunity. With the goal of building a truly sustainable, accessible, and multimodal transportation system, funded projects include:

- A study of Ala Moana parking, in collaboration with Columbia University and the University of Hawaii at Manoa, to determine how to better manage more than 32,000 off-street parking spaces and 1,300 street parking in order to aid drivers and advise local municipalities against funding and building additional parking supply.
- An EV charging station incentive program with Hawaii Energy to provide up to $5,000 for a newly installed EV charger at workplaces or multiunit dwellings. The program is designed to increase Hawaii's EV charging network and to spur EV growth locally.
- A bikeshare program, Biki, with partner Bikeshare Hawaii, to launch a new system with 1,300 bikes and 130 self-service stops that provides the public with high-quality, convenient, reliable, and affordable bikeshare services. With grant support originally provided through the Ulupono Fund at the Hawaii Community Foundation, Biki works to enhance community health and livability while strengthening already established public transportation systems.
- A report analyzing the annual costs of the vehicle economy in Hawaii, which provides comprehensive data to local leaders, weighing expenditures and investments to best address community needs as well as help residents to make informed choices for themselves and their families about the costs they bear directly as consumers and taxpayers.

Over the next several years, Ulupono Initiative is committed to executing targeted efforts to support clean transportation initiatives that right-sized parking policies and practices, build capacity for multimodal solutions, empower EV and AV outreach and acceptance, and support local and state goals around VMT.

About the Team

Development of Hawaii's A2CES framework was led by Ulupono Initiative and Me3—with Kathleen Rooney, director of transportation policy and programs at Ulupono Initiative, and Kelly Coyner, Founder and CEO of Me3—at the helm. As a result of the framework, Rooney also served on Hawaii's Autonomous Vehicles Legal Preparation Task Force to prepare the state with laws and regulations required for AVs by examining the adaptation and testing of the technology and existing laws relating to legal and insurance regulation of AVs and making recommendations for AVs in Hawaii. Ulupono Initiative was founded in 2009 to improve the quality of life for the people of Hawaii by working toward sustainable solutions that support and promote locally produced food, renewable energy, clean transportation, and better management of water and waste. Me3 works with communities like those served by Ulupono to deploy automated, connected, and electric mobility systems.

Over the next decade, statewide adoption of AVs in combination with an integrated multimodal network will allow Hawaii residents to transition from spending hours commuting on congested highways and roads to safely riding in autonomous electric cars, buses, and transit. Simply put, AVs must be electric and shared to realize their full potential to drive renewable energy goals and give those who cannot drive more autonomy through mobility without contributing to ever-increased traffic by increasing the number of cars on the road and trips made by them.

Kathleen (Katie) Rooney is Ulupono Initiative's lead on transportation-related policy and programs in the advancement of cleaner, multimodal transportation in Hawaii and reducing our dependency on cars. Her current projects include right-sizing parking policies and helping expand immediate transportation choices and access to those choices. She brings 15 years of national experience in transportation and planning, combining both to advance community visions and goals in many diverse communities across the nation. For example, using a combination of analytics and storytelling, she helped two West Virginia towns strategize around health, leading to two trail program implementation grants. She also helped to conceptualize a resiliency center in Washington State, operationalize multimodal accessibility metrics in Florida, and develop a statewide transportation demand management framework for New York State. Prior to joining the Ulupono team, Rooney served as a project manager of Renaissance Planning in Orlando, Florida, and as a senior manager at ICF International in Washington, DC. Katie holds a master's degree in public policy from the University of Maryland and a bachelor's degree in political science from Tulane University in New Orleans.

Kelley Coyner co-founded Me3, now known as Innovation4Mobility, to bring accessible automated connected electric and shared mobility systems to communities and campuses. She helped build several emerging businesses to bring A2CES to scale. Now she forges partnerships between technology firms and cities and campuses to bring mobility to all. As a two-time CEO and member of the 2008 Presidential Transition Team, she has fostered innovation needed to rebuild after devastating natural disasters, terrorism events, and financial crises. Coyner has advised more than 70 nonprofit boards globally, served as Mobility Innovation Lead for Stantec and Executive Director of the Northern Virginia Transportation Commission, been confirmed as head of the Research and Special Program Administration, and held appointments at MIT, Harvard, the Coast Guard Academy, the National Academy of Sciences, the Volpe National Transportation Systems Center, and George Mason University. In addition to the A2CES Framework for Hawaii, she has authored mobility innovation frameworks for a dozen cities, states, and private owners. In January 2021, the National Academies Press published her assessment of low-speed automation in public transportation, and she served as the principal investigator for comprehensive assessment of research needs in AVs and shared mobility across nine categories: safety, freight, land use, equity, social impacts, transit, data, infrastructure enablers, and planning and modeling. Her nom de plume is Mobility Mama, and she is a published travel writer focused on multimodal family adventure travel by microbuses and by foot on Incan roads, biking, fishing boats, trains, and planes. She is a disruptive tech whisperer working to create equitable and sustainable communities.

Climate Preparedness and Resilience Exercise Series

Emily Wasley, Corporate Climate Risk, Adaptation, and Resilience Practice Leader at WSP, President of the American Society of Adaptation Professionals (ASAP), former Adaptation Science and Informing Decisions Program Manager with the U.S. Global Change Research Program (USGCRP)
Washington DC, Houston-Galveston Region, Texas

Background

Planning for climate adaptation represents an important form of hazard mitigation planning (Figure 9.2). In 2013, federal government departments and agencies began to build agency adaptation plans to strengthen hazard mitigation efforts related to climate risks. In a partnership between the Office of Science and Technology Policy, the Obama Administration's White House Council on Environmental Quality, and FEMA's National Exercise Program, the team worked interdepartmentally to design the Climate Change Preparedness and Resilience Exercise Series in five jurisdictions across the nation.

The five areas selected include:

1. Hampton Roads Area, Virginia.
2. Houston–Galveston, Texas.
3. Salt Lake County, Utah.
4. State of Colorado.
5. State of Alaska.

FIGURE 9.2 Aerial view of flooding caused by Hurricane Harvey.

The USGCRP supported five regional workshops in 2014 as part of this Climate Change Preparedness and Resilience Exercise Series. For each workshop, location-specific brochures describing observed climate trends and projected future climate conditions were developed using information from the 2014 National Climate Assessment and other peer-reviewed science at the regional and local scale.

The intent of these pilots was to bring together "whole community members" across federal agencies and local communities to assess and plan for their region-specific vulnerabilities and interdependencies associated with the impacts of climate change. This effort was intended to advance collaborative climate change preparedness and resilience planning on the ground and help create models for other communities and agencies to follow.

Houston–Galveston, TX, Climate Change Preparedness and Resilience Exercise

The Houston–Galveston region was identified as a key location to perform the exercise because this region experiences intense hurricanes, is a key supply chain hub through its shipping channel and port, and is home to the Texas Medical Center, which is the world's largest health-care center. The Texas Gulf Coast averages approximately three tropical storms or hurricanes every four years (at the time of the exercise). These trends indicate that there will be coastal storm surges that could bring heavy rainfall and damaging winds hundreds of miles inland. An increase in expected sea-level rise will result in greater damage from storm surges along the Gulf Coast.

As of the 2018 data, the relative sea-level rise along the Houston-Galveston area and the Texas Gulf Coast was twice the global average. A single storm surge in Galveston Bay would threaten much of the U.S. petroleum and natural gas and refining capacity.

These pilots were sponsored by the White House National Security Council Staff, the Council on Environmental Quality, and the Office of Science and Technology Policy. In collaboration with the National Exercise Division and hosted by the City of Houston, Mayor Annise Parker, and The National Aeronautics and Space Administration (NASA)'s Johnson Space Center, the pilot brought together a diverse group of experts looking to plan for the worst, develop a regional resilience plan, and see where gaps existed that could be addressed to make sure the worst-case scenario was never realized.

Goals and Objectives

This effort was set for the Houston-Galveston area to identify and refine climate change preparedness and resilience requirements and initiatives, in collaboration with critical whole community stakeholders. The purpose of this work was to advance the climate adaptation dialogue and identify collaborative and sustainable approaches to community-based climate preparedness and resilience capabilities. Workshop objectives included:

1. Examined methods to better integrate existing and emerging scientific information and other requirements into current and future planning to manage and adapt to climate risks and vulnerabilities.

2. Identified collaborative and sustainable whole community approaches to advance and sustain local climate preparedness and resilience programs, policies, and strategies.
3. Examined investment opportunities and the development of coalitions between local, state, federal, and private sector partners to support climate preparedness and resilience.
4. Examined relevant effects of climate change and hazard mitigation strategies for populations of disproportionate impact (vulnerable communities and populations).

The workshop focused on the following outcomes:

1. Improved collaboration with and between whole community partners on climate preparedness and resilience strategies.
2. Identification of new research, information, and capabilities that will support local preparedness, adaptation, and hazard mitigation planning.

Workshop outputs included the following:

1. Workshop Summary Report that addresses key discussion points and identifies climate preparedness and resilience information, innovations, and initiatives.
2. Potential climate change risks and vulnerabilities to be addressed in the local, state, and regional Threat and Hazard Identification Risk Assessment processes.

Approach and Challenges

Conducted on Monday, October 6, 2014, the Houston Climate Change Preparedness and Resilience Exercise brought together more than 90 whole community participants across the Houston–Galveston region to conduct a comprehensive and thought-provoking conversation on climate change adaptation and mitigation and what solutions could be realized to plan for resilience. The exercise was designed based on sound science portraying Houston under a "business as usual" high-emissions scenario with higher annual average temperatures, changing precipitation patterns, and rising seas, layering on an acute event: the increased risks from a hurricane's storm surge and heat waves in the mid-twenty-first century.

Using climate science from the Third U.S. National Climate Assessment as a foundation supplemented by science from local experts, discussions focused on the effects of climate change and extreme weather, the associated challenges and opportunities the Houston–Galveston region may face, and specific immediate actions to collaboratively and sustainably prepare, plan for, or help mitigate these future projected climate impacts to critical infrastructure.

Exercise participants included local, state, and federal representatives, as well as the private sector, nongovernment, and academic partners who have roles, responsibilities, and expertise as they relate to climate adaptation, hazard mitigation, and resilience planning efforts (Table 9.1).

TABLE 9.1 Descriptions for the core capabilities examined during the workshop.[9]

Core capability	Description
Community Resilience	Lead the integrated effort to recognize, understand, communicate, plan, and address risks so that the community can develop a set of actions to accomplish mitigation and improve resilience.
Long-Term Vulnerability Reduction	Build and sustain resilient systems, communities, and critical infrastructure and key resources lifelines so as to rescue their vulnerability to natural, technological, and human-caused incidents by lessening the likelihood, severity, and duration of the adverse consequences related to these incidents.
Operational Coordination	Establish and maintain a unified and coordinated operational structure and process that appropriately integrates all critical stakeholders and supports the execution of core capabilities.
Planning	Conduct a systematic process engaged in the whole community as appropriate in the development of executable, strategic, operational, and/or community-based approaches to meet defined objectives.
Risk and Disaster Assessment	Assess risk and disaster resilience so that decision-makers, responders, and community members can take informed action to reduce their entity's risk and increase their resilience.

Reprinted with permission from National Preparedness Goal, September 2011. US Homeland Security.

Results

Feedback from participants was overwhelmingly positive. They appreciated the opportunity to discuss climate change impacts and adaptation opportunities and challenges and, importantly, to build networks to take action toward climate change resilience planning in the Houston–Galveston region. Participants from private sector organizations and nongovernment organizations highlighted ongoing, as well as projected, activities and opportunities in support of the below efforts (Figure 9.3):

- Identification of opportunities to align natural and built systems:
  - Participants noted the need to adapt infrastructure design standards to address projected consequences and cascading effects from climate change (particular to sea-level rise and higher temperatures) and discussed opportunities to achieve better alignment of natural and built/engineered infrastructure and systems.
- Recognition of critical interdependent lifelines:
  - Participants suggested mapping critical lifelines (e.g., electricity, telecommunications, water, and transportation) across the region to demonstrate interdependencies and identify areas that are most vulnerable. Participants from academic institutions and federal agencies offered modeling resources to support these efforts.
- A need for strategic messaging:
  - Participants from all organizations discussed the importance of strategic messaging to drive action now to reduce long-term vulnerability and prioritize investments in large-scale resiliency projects in order to save money later.

---

[9] National Preparedness Goal, September 2011. US Homeland Security.

FIGURE 9.3 Water coming over the streets in Kemah during Hurricane Harvey.

- Many partnerships exist, but a more coordinated approach is needed:
  - Participants identified a wide variety of existing coalitions, partnerships, associations, and committees in the Houston–Galveston region that address sustainability, resiliency, and preparedness issues; however, participants acknowledged the need to unify disparate efforts being undertaken into a more cohesive climate change resiliency governance framework.
- Agreement from all participants on the need for a unified regional resilience framework:
  - All participants verbally agreed to pursue a Houston–Galveston Regional Resilience Plan that involves the whole community and contains flexible and dynamic short-term requirements driving toward a long-term strategic vision that "builds a better, more resilient Houston–Galveston region."

Additional Commentary

An excerpt from Wasley's 2019 American School in London high school commencement speech:
Acts of courage have the power to change our lives. At some point in life, every one of us either has, or will, face an insurmountable setback. Whether it be the death of a close friend or family member, the loss of a job, a breakup or divorce, personal illness, or a major disaster, if we can see these setbacks as chances to be brave, then we open ourselves up to opportunities to thrive and grow. It is at these times that we come to know ourselves because we begin to understand how deep our reservoir of resilience runs, and we emerge from crises stronger.

Working in the climate field has been and will continue to be a constant battle: but it is a battle that has taught me the true value of courage and connection. I now have the courage to stand up to people who say they do not believe in climate change, as if it is a religion instead of a scientific fact. I have the courage to understand that I cannot heal the world on my own—that it requires everyone's help. And I have the courage to be honest with people about what I know about the fate of our planet as a result of a changing climate.

And, unfortunately, it does not paint a pretty picture.

We are on the brink of social and ecological collapse triggered by the climate catastrophe caused by humans. Our global population has exploded having almost tripled from 2.5 billion in 1950 to 7.7 billion today. We have escalated urbanization with more than half of the world's population now living in towns and cities. With more people, we have higher demands for food and water, creating significant economic insecurities and resource inequalities worldwide. Since the 1950s, we have increased our consumption of Earth's natural resources tenfold. As we continue to deplete these resources while emitting toxins into our atmosphere, we are eroding the very foundations of our economies, livelihoods, food security, health, and quality of life worldwide. We are already experiencing suffering, trauma, and human migration because of these changes.

As a climate steward, I witness the suffering of our planet and the frontline communities who are hit first and worst on a regular basis. One of my specialties is known as "scenario planning"—where I use information that I just laid out for you about the state of our planet, add in hypothetical cascading impacts, and design scenarios to help my clients visualize potential future conditions. Basically, I get paid to "future trip," and at times, I design some of the worst possible futures—futures that I hope never come true. However, during my career some already have.

In 2014, I worked on President Obama's Preparedness Pilots. These pilots took place across several U.S. cities with the goal to assess and prepare for climate risks through scenario planning. One scenario I designed was set to play out 30 years from now and was based on scientific data from the latest climate models. This particular scenario was designed to inform the development of a regional resilience plan for the Houston-Galveston area—a plan that for various reasons—never came to fruition.

Three years later, in the fall of 2017, this hypothetical scenario I had designed played out in real time when not one, but three powerful hurricanes hit the Gulf Coast of the United States in succession. Hurricane Harvey pummeled Houston for four days dumping more than 40 inches of rain, flooding thousands of homes, and displacing over 30,000 residents. Harvey alone cost $125 billion in damages. As the storm ravaged the region, I watched in horror, knowing that the futuristic scenario I had designed was unraveling much sooner and with far more force than anyone expected.

The people of Houston displaced by Hurricane Harvey were some of the same people displaced by Hurricane Katrina in 2005 (Figure 9.4). They continue to live on the frontlines of the systemic injustices we face today. They have experienced not only the acute trauma of living through these extreme weather events, but they have been forced to relocate their families, their livelihoods, and their communities to areas that are not familiar, areas that may not welcome them because of their skin color. This acute trauma can lead to chronic suffering because they no longer have a sense of belonging and are forced to find, yet again, their sense of purpose.

FIGURE 9.4 High and fast water rising in Bayou River with downtown Houston in the background under a cloudy blue sky. Heavy rains from Category 4 Hurricane Harvey caused many flooded areas in the greater Houston area.

When I think about the scenario I designed for the Houston–Galveston area, knowing that more could and should have been done to enhance the resilience of this region, I am left heartbroken. But these are experiences where I get to sit with the pain, process it, learn from it, and then do everything in my power to be of service through my climate work.

About the Team

Emily Wasley is a climate change adaptation professional passionate about empowering everyone to thrive in the face of uncertainty. She has demonstrated experience convening diverse, whole community members to understand and assess risks and codesign solutions to address climate change. She is an adaptive thought leader skilled in connecting climate science with future scenario planning and strategic foresight to prepare people, businesses, and communities to be Future Ready™. She has more than 16 years of experience in the climate change, sustainability, adaptation, and resilience fields helping government entities, water utilities, and Fortune 50 to 500 corporations institutionalize climate considerations into existing risk management, business continuity, and sustainability efforts. She leads WSP's corporate climate risk, adaptation, and resilience practice supporting clients in identifying, assessing, and managing climate-related risks and opportunities, the various risks and opportunities in alignment with the Task Force on Climate-related Financial Disclosures recommendations. Wasley serves as Board President for the ASAP, on the Steering Committee of Women in Climate Tech, and as West Coast Future Ready™ Advisor for WSP USA.

10

Talent and Education

Zivica Kerkez/Shutterstock.com.

> The biggest message I have for young women is, don't start cutting off branches of your career tree unnecessarily early. Sometimes women say, "I know I want to have a family or play in the local symphony," and then they start pulling themselves out of their career path. You don't have to take yourself out of the running before you even start.
>
> —Mary Barra, Chairman and Chief Executive Officer, General Motors

By now you should have an understanding of the significant gender imbalance within the automotive, transportation, and mobility industries—especially in technical roles. Now let us take a look into where this issue begins.

If we take a step back and look at what is happening in K-12 education, data shows that female student achievement in mathematics and science is on par with their male peers, and female students participate in high-level mathematics and science courses at similar rates as their male peers.[1]

However, things go awry if you examine higher education. Data shows the rates of science and engineering course enrollment for women shift at the undergraduate level, and gender disparities begin to emerge, especially for minority women. Although

---

[1] https://ngcproject.org/statistics

interest in STEM studies has increased in the last 10 years, more than 32% of women make the decision to switch out of STEM degree programs in college prior to graduation.[2] By the end of university, the numbers are even more discouraging: In 2018, only 21% of bachelor's degrees in engineering were awarded to women.[2]

To make things worse, women are leaving STEM careers at an alarming rate. Only 30% of women who earn bachelor's degrees in engineering still work in the field 20 years later. Of those who left engineering professions, 30% cite their organization's culture as the reason for their career change.

How can we combat this trend? How can we attract more gender diversity to the field, and how can we retain them for the duration of their careers? For starters, we have to create an environment that is inclusive, welcoming, and appealing to a diverse workforce. Second, regardless of gender, we have to adequately prepare today's engineers for tomorrow's industry, which will allow top talent to see career path opportunities and help increase industry retention rates.

The Society of Women Engineers (SWE) has programs at all levels of the education pipeline to increase interest in STEM studies among students of all ages. SWE also provides support for professionals throughout their careers. Additionally, organizations like the Center for Automotive Diversity, Inclusion, and Advancement (CADIA)—which is featured in this chapter—are doing the work today to provide diversity, equity, and inclusion (DEI) tools, networks, insights, and practical advice to industry companies with the immediate goal of increasing diverse leadership within the industry.

Finally, the industry has to focus on ensuring the workforce has the right skills for today's industry and is prepared for the technologies and principles that are part of the new mobility industry. Organizations like The NEXT Education—which is also featured in this chapter—and the Michigan Mobility Institute, co-founded by Jessica Robinson, are creating curriculum and tools that boost readiness for high-demand, high-wage careers in mobility tech, intelligent transportation, autonomy, electrification, engineering, software development, cybersecurity, and data sciences.

The stories featured in this chapter exemplify the extraordinary work that is being done to ensure not only that the mobility workforce is more diverse—in every sense of the word—but is also prepared to enter a new era of transportation that is vastly different than anything this industry has seen before. Additionally, these stories will all share insights into what needs to be done to make certain our future talent pipeline has the right resources and is sustainable for future growth in the sector.

The final case study showcases a student group and what happens when you empower young people to envision the way the world should move.

Championing Diversity, Equity, and Inclusion across the Industry

Cheryl Thompson, CEO and Founder of the Center for Automotive Diversity, Inclusion, and Advancement (CADIA)

Margaret Baxter, Executive Director, Center for Automotive Diversity, Inclusion, and Advancement (CADIA)

---

[2] https://research.swe.org/wp-content/uploads/2020/11/FOR-PRINT-SWE-Fast-Facts_Oct-2020.pdf

Background

Historically, the automotive and mobility industry has not been known to be the most diverse industry, and the numbers of diverse talent get smaller in leadership and executive levels.

DEI initiatives have taken center stage among larger companies such as the automakers. Those committed to DEI are actively working to create meaningful change among underrepresented groups within the workplace. These types of initiatives are vital to creating and maintaining a successful workplace. DEI initiatives work to ensure all people can thrive personally and professionally while also bringing new and differing perspectives to the workplace, which can spur ideas and innovation.

The case for change is strong. According to McKinsey's "Diversity Wins: How Inclusion Matters" report, companies with gender, racial, and ethnic diversity perform better, are more profitable, and are more successful in retaining top talent.[3] If we look at talent in the United States and Canada, we can see that our demographics are shifting. According to an NPR report, U.S. Census Bureau data showed that starting in 2020, 50.2% of the publication under the age of 18 came from an ethnic group or minority race.[4] Additionally, according to the Deloitte Millennial Study, Millennials and Gen Z are looking for workplaces that are inclusive, have flexibility, and are places in which they can make a meaningful contribution.[5] To attract, retain, and develop the best talent, the industry must double down on DEI commitments and position itself as inclusive and diverse.

In recent years, automakers have made progress on their DEI journey, but the work has been slow. On the heels of the industry giants, the supplier network and other smaller organizations are just getting started.

CADIA was founded in 2018 to accelerate the DEI work being done. The nonprofit's goal is to move the entire industry forward in terms of creating inclusive cultures where everyone feels welcomed, valued, seen, and heard in order to attract, retain, and advance the very best talent.

Goals and Objectives

CADIA's guiding mission is to double the number of diverse leaders in automotive and mobility by the year 2030. The organization serves in a consultative and supportive capacity to automotive and mobility companies, an especially compelling offer for those that may not have their own internal DEI teams and added support for those that do. Along with industry leadership, CADIA aims to:

- Drive Systemic Change:
    - Collect and share best practices and trends on talent system redesign, equity, inclusion, and employee engagement.

---

[3] https://www.mckinsey.com/featured-insights/diversity-and-inclusion/diversity-wins-how-inclusion-matters
[4] https://www.npr.org/sections/ed/2016/07/01/484325664/babies-of-color-are-now-the-majority-census-says
[5] https://www2.deloitte.com/tr/en/pages/about-deloitte/articles/millennialsurvey-2018.html

- Develop a comprehensive Speakers Bureau on DEI initiatives, value proposition, and state of DEI in the automotive industry.
- Create a DIE Barometer™—an annual survey illustrating the automotive industry's DEI journey.
- Host a roundtable series that brings DEI champions together to share best practices and learning, corporate workshops, and training, as well as provide one-on-one coaching.
- Support Leadership Commitment:
  - Host a workshop for leadership teams to bring DEI into sharp focus as organizational leaders.
  - Offer assessments that provide a snapshot of where a company is on its journey, with a roadmap for go-forward recommendations.
  - Provide quarterly check-ins with the leadership team and/or DEI champion(s) and offer strategic advising.
- Champion Diverse Talent:
  - Educate others through events, low-cost training, and workforce development activities through the CADIA Academy.
  - Host career fairs to bring diverse talent and forward-looking employers together, as well as corporate workshops and coaching.

Approach and Challenges

Many companies have been trying to solve the lack of progress in DEI inside a vacuum. To address this, CADIA convened leaders and stakeholders from across the industry to understand the culture and shared complexities of the industry and to collaborate, share best practices, and "crowd-source" effective solutions.

An exploratory meeting was held where DEI champions from various automotive companies gathered to discuss best practices, share open conversations on challenges and opportunities, and learn from one another in a peer environment.

Next Steps

Drawing on the success of the first meeting, there was enough interest to make the CADIA DEI Roundtable Series a formal offering. Co-chaired by two women from the industry, the Roundtable took off quickly. Currently, industry leaders and stakeholders meet monthly. During these sessions, participants have garnered enormous value in being able to learn from one another.

Going forward, CADIA members will apply the learnings and leverage DEI initiatives within their companies as a business strategy enabler to deliver on the mission of the organization and achieve business objectives. The group has doubled in size over the last year and will continue to expand to reach automotive companies throughout North America and the globe. CADIA is currently evaluating growth plans to include the retail automotive industry and has started an initiative with a dozen automotive CEOs who have banded together to accelerate change within their companies and to hold one another accountable.

About the Team

CADIA was founded by Cheryl Thompson and supported by Margaret Baxter as Executive Director, pulling from their diverse careers in the automotive industry including engineering, manufacturing, and legal, to bring a well-rounded perspective in service of helping the industry succeed with DEI initiatives.

Cheryl Thompson is the founder of CADIA. CADIA supports DEI for the Automotive Industry by providing professional development for individuals, along with resources, programs, and tools that drive organizational evolution.

A veteran of the automotive industry, Thompson has more than 30 years of experience at Ford Motor Company and American Axle and Manufacturing in positions ranging from skilled trades, operations, engineering, and global leadership.

She is trained in diversity and inclusion and career and leadership coaching and is Six Sigma trained and certified as a Black Belt. Thompson has been recognized as a 2019 Influential Women in Manufacturing Honoree, a 2019 Corp! Magazine Salute to Diversity award winner, and Marketing and Sales Executives of Detroit Platinum Awardee and is the recipient of two Diversity and Inclusion Awards from Ford Motor Company.

A sought-after voice and speaker in the automotive, manufacturing, aerospace, and defense industries, Thompson has performed keynote addresses, workshops, and breakout sessions for a number of companies and events, including TEDxWindsor, Women in Manufacturing, the American Automotive Summit, and SWE.

Margaret Baxter is the Executive Director for CADIA. She runs CADIA's day-to-day operations and helps develop tools, offerings, and resources to accelerate the pace of diversity, equity, and inclusion in the automotive industry.

She is trained in diversity and inclusion, has conducted trainings and workshops, is an experienced moderator, and has spoken at numerous industry events. Having worked in the automotive, legal, government, and nonprofit worlds in many different capacities, Baxter brings a dynamic background to CADIA.

Designing a Mobile Future

Maria-Luisa Rossi, Chair and Professor, MFA Systems Design Thinking, CCS Graduate Program at College for Creative Studies
Detroit, Michigan

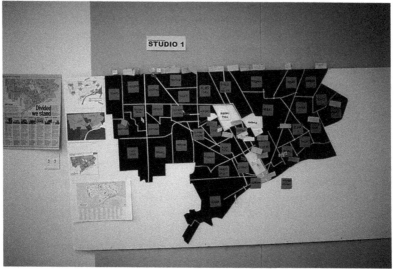

Background

As a part of a citywide effort to rethink equitable development, Detroit Equitable Mobility 2030 explored innovative neighborhood mobility proposals with College for Creative Studies (CCS). This effort was made in partnership with Design Core, the steward of Detroit's UNESCO City of Design designation, and an organization serving the creative community by championing design-driven businesses through their journey as critical members of Detroit's economy. Design Core convened the initiative to activate Detroit's commitment to using design as a driver for equitable development and prosperity. The CCS Master Program Systems Thinking Design (formerly known as Integrated Design) worked to understand how human-centered design research and collaborative partnerships can lead to a more "mobile" future for Detroiters by 2030. The two-semester student project explored innovative neighborhood mobility proposals as part of a citywide effort to apply user-centered design as a driver for equitable development.

Knowing that getting mobility right could be a substantial advantage for cities, this team started by analyzing the role of transport planning and the significant mobility impacts on habitats.

Acknowledging that mobility is one of the people's daily life experiences regardless of their socioeconomic status, we have valued an inclusive perspective on Detroit, a city with vast and diverse human needs, and let these factors shape the vision.

Goals and Objectives

The main goal was to drive growth and urban well-being. This was done by identifying insights and building valued proposals associated with mobility allowing for engagement, accessibility, and equitability. The initiative objectives sought to leverage academic, industry, government, and community partners to develop limitless ideas to increase people's ability to move freely and safely in and around their communities in Detroit. This project offered students a unique opportunity to build a shared vision of the future of mobility through a holistic and integrated approach via scenario design and service design courses.

Approach and Challenges

Considering the many uncertainties surrounding Detroit's future of mobility, scenarios needed to consider plausible alternative paths for field stakeholders to choose the best path forward. Specifically, the project focused on participatory scenarios, a bottom-up participatory planning design with the ability to integrate local perspectives into development strategies. The Scenario Design outcome was obtained with a methodology comprehensive research study; an in-depth, data-driven report study; expert opinions from the community, academia, and industry; as well as user-centered and cocreation sessions.

The research study was conducted with three "descriptors" or variables of interest in mind: health, employment, and neighborhood. Talks with experts were organized, one for each variable of interest. The variables were later identified as placemaking, entry point, and care exchanging. Subject-matter experts from city government, academia, nonprofit organizations, community representatives, activists, and consulting firms shared their considerable substantive experience in a specific field and asked for projections for the descriptor for 2030, along with their assumptions.

As for the scenarios, each was broken down into respective approaches to tackle certain social and environmental challenges that may be experienced in various communities and by different populations. First, "placemaking" was focused on how Detroit youth can build a sense of community ownership. The "entry point" scenario explored entrepreneurship and training to address a lack of entry-level jobs in the City of Detroit and, to take it a step further, to address the lack of training opportunities for Detroiters. The "care exchanging" scenario offered creative community approaches to public health, especially regarding diabetes and obesity, and barriers that exist and keep Detroiters from accessing meaningful health care. Placemaking, entry point, and care-exchanging scenarios invited the students to consider the overall scenario and prepare to advise on alternatives and viable solutions given the circumstances.

In the second phase of the project, the students worked with Detroit nonprofit Focus: HOPE, cocreating to identify community needs and priorities to validate the services and proposed solutions.

A Guidance Team was also established to help students to design the overall scenario framework and then advise on the formulation of the alternative scenario storyline. This complex organization ensured that the students heard all entries. Codesign activities took place throughout the process in order to build the services with the users. This level of engagement offered additional opportunities for people to be creative, share insights, and visualize what they wanted their community to look like. Special attention and consideration was given to individuals who are often excluded from development processes like this. The students were able to especially interpret this cocreation process because they were able to work directly with community stakeholders to absorb the emotional context of Detroit's past, current, and future environment.

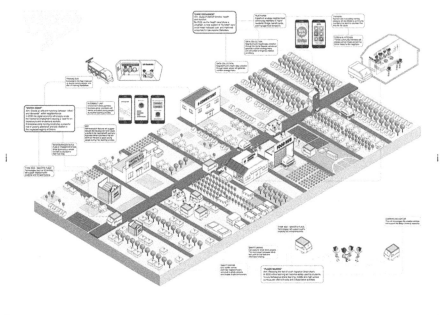

Reprinted with permission from © MFA System Design Thinking.

Scenario Design Details

- *Placemaking:* The scenario aims to support the motivation and abilities of Detroit's young people to participate in their community development to enable safety and security at the neighborhood scale and reduce the risk of youth migration. By seeking to understand the emotional distress of Detroit youth, the students were able to consider how isolation and a sense of belonging are key themes that need to be considered when exploring how to activate the next generation of community leaders.

  In 2030, following the current trend of STEM education, technology will become a vital component of the grade school curriculum and online learning will become widely used by students. To counterbalance the trend of online learning, middle and high school curricula will need to offer soft skills training so that students are trained to excel with teamwork, adaptability, and creativity, especially as it relates to cultural awareness. Considering the new digital learning landscape, additional efforts will need to be made in order to connect students. A "Learning Incubator Unit" was included in this design to encourage creative work and help create participation opportunities by creating a mobile, physical environment for this creativity and collaboration to exist.

- *Entry Point:* This scenario challenge explored the gap and discrepancy between the scarcity of Detroit's entry-level jobs and the public perception of skills necessary to apply or be eligible for entry-level jobs. Counter to the story of pervasive youth unemployment, Detroit young adults represent the hope for the city. In 2030, the focus of capital investment is expected to continue to move from human labor and talent development toward automation technologies. The digital economy will sharply erode the traditional employer-employee relationship and will create a need for individual entrepreneurs and on-demand workers. New nonformal education and experiential learning models could be useful in an era of greater individual entrepreneurship. The aim is to create an efficient matching between "offers" and "demands" that create jobs and wealth within neighborhoods. The proposed tailored approach will be delivered in phases. Self-evaluation, behavioral testing, games, and other simulations may offer a bridge to help young people step into their adult lives with more motivation. These self-evaluation profiles could be matched with partners/businesses to find appropriate volunteer, employment, and training opportunities.

- *Care Exchange:* The Care Exchange Scenario focused on public health concerns, specifically looking at implications and challenges associated with high rates of child and adult obesity and diabetes among Detroiters. In the United States, health-care costs are approaching 18% of GDP. This unsustainable system suggests the need for a more efficient health-care system. This new approach should integrate a robust primary care system (easy access to family physicians in the community) with strong incentives to keep patients healthy and well (from hospital to home). By 2030, the fourth industrial revolution will ensure that humans live longer and healthier lives. The use of genetic information will push the use of "designer drugs" specifically created to fight disease and will increase the use of screening. Real-world data collection will increase accuracy and allow us to predict which patients are most likely to need follow-up care or preventive tips to stay healthy. As the health-care future is uncertain, this new system of bundled care could lead to reduced costs and improved outcomes for low-income Detroiters if neighborhood community members all contribute toward the common goal. Included in this bundled care system:

  - Nutrition and food safety training sessions will be offered to community members who want to volunteer their time for the cause.

  - Neighborhood health data collection (via sensor and vehicle design) to generate nutrition strategy and menu. A platform like this can enable neighborhood community members to "bank" residents' food stamps and manage food donations; trained community members can prepare healthy meals for delivery and distribution.

Results

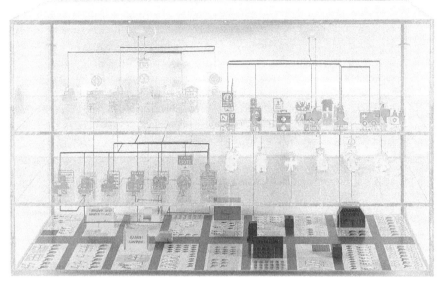

Reprinted with permission from © MFA System Design Thinking.

The autonomous mobile pods included in this design bring the twenty-first-century curriculum directly to youth in each neighborhood, removing elements of barriers such as transportation or broadband access. Larger mobile units could transform daily commutes into tailored place-based entrepreneurship training and mentorship for the neighborhood's young adults. Finally, mobile health units collect relevant data to generate nutrition plans and menus, fostering healthy communities in which health-care barriers are dissolved through the deployment of localized and mobile health-care facilities. The drawing illustrates three long-term equitable mobility design scenarios: Entry Point, Placemaking, and Care Exchange. The design scenarios aim to improve the work prospects, youth retention rates, and health outcomes for underserved Detroit neighborhoods.

Building on "Detroit Equitable Mobility 2030," Scenario Design students developed services offering a systematic design process that comes from the perspective of the user. The service design of the placemaking scenario was soft-skill education services. These services would provide access to a new education system through the learning incubators with mobility connections. These services would provide support for users to access community collaboration and networking opportunities, creating additional paths to meaningful employment.

The service design of the care exchange scenario focused on food access, a key component of a healthy lifestyle, as well as medical access, comfortability, and senior care. The care exchange platform would enable the neighborhood community members to "bank" their time and provide nutrition and food-safety training sessions to community members who volunteer. The health data collection service would be used to develop programs and add services aimed at decreasing the number of chronic illnesses and emergency care visits, collecting neighborhood health data, and providing door-to-door care for low-income Detroiters.

The students' long-term vision of the desirable future mobility is one that is equitable and provides accessibility for improved quality of life through healthy communities, nutritional access, educational and training access, and community ownership.

This project was born with caution and thoughtfulness as a program that made neither mobility nor transportation a focus. Being in an area of absolute experimentation, we were free to approach mobility from the perspective of a user who is not listened to but is penalized enormously by the lack of transport. In its early stages of organizing an exhibition on innovations in mobility, the Cooper Hewitt Smithsonian Design Museum contacted CCS and selected the two consecutive projects' elements, scenario and service design.

Organized by Cara McCarty, the museum's director of curatorial; Cynthia E. Smith, curator of socially responsible design; and Julie Pastor, curatorial assistant, and featuring 40 projects, " The Road Ahead: Reimagining Mobility" exhibition explored salient topics around the future of mobility and the urban environment, including accessibility, equity, trust, safety, and security. These challenges presented a critical opportunity to pursue a new user-centered vision for streets and infrastructure to create more livable, inclusive, and equitable cities. Equitable Mobility, Digital Divide, and Climate Change are the Systems Design Thinking program's topics for forward-looking and inclusive projects to enhance people's physical and psychological assets.

The academic leadership at CCS believes that an inclusive design approach is a way of working that helps designers listen, learn, and adjust their lenses and biases to bring the full spectrum of diversity into the design process.

About the Team

Maria-Luisa Rossi's work at the CCS Graduate Studies brings her entrepreneurial, globally focused, and empathetic cultural approaches to the next generation of designers. She focuses on the seamless capacity to deal with the tangible and intangible aspects of people's experiences. At CCS, she is preparing "facilitators" capable of addressing global and local challenges, focusing on social innovation. Her projects are concentrated on research, cocreation, and people-centered processes.

In 2008, her interdisciplinary attitude, design and brand strategy knowledge, and business acumen brought her to the team that launched the new Graduate Program in Design at CCS in Detroit. In that capacity, she has developed the curriculum and set standards of excellence for MFA Systems Design Thinking, validated by a successful alumni job placement in corporations and design consultancies setting.

Rossi's professional career has been independent and international. She attended the premiere master's program in industrial design at the Domus Academy in Milan, thanks to a European Scholarship she won by designing the first wearable computer and winning a design competition. The project was featured in the prestigious *Domus* magazine and gave her visibility around Europe and in the design world. The wearable computer project, "The Walking Office," can be found in the Henry Ford Museum Permanent Design Collection.

Following her studies, she founded the design consultancy Iavicoli and Rossi, working in areas varying from interior architecture to tableware.

Rossi is a Detroit City of Design Research Lab Advisory Board member working on the vision for inclusive growth and providing guidance on the overall framework for the lab. Rossi has conducted workshops and lectures in Singapore, LA, Mexico City, Istanbul, Ankara, São Paulo, Shanghai, Gratz, Brasilia, and Taiwan.

Her most recent lectures are "Ladies Design Salon," about city systems and inclusive futures; "Italian Design Day in the World," addressing the importance of design on the shaping of future cities; and "International Seminar on Creative Economy Creative District," about creativity and design as an asset of creative tourism.

Preparing the Workforce for the Next Generation of Mobility

Elaina Farnsworth, CEO at The NEXT Education
Michigan

Background

The world is connected in a way that it has never been before and today's advancements in transportation—connected cars, AVs, and unmanned aerial vehicles (UAV)—are pushing the limits of innovation. Our world is changing at lightning speed. These advancements in technology and information have caused a significant transformation for our world today and for our workforce of the future.

Who will service a network of AVs on the ground and in the sky? Who will secure and maintain a connected infrastructure to ensure the safety and efficiency of next-generation vehicles? How will today's workforce adapt to these changes? Today, there is no easy solution to those questions.

To move forward as an industry, we need to close the growing talent gap and train people in the concepts of new mobility, smart cities, and intelligent infrastructure. It is impractical today to find an entirely new team to service these new concepts. Not only can companies not justify a complete employee turnover, but they also cannot afford it. Fortunately, the industry has a large pipeline of talent already in its midst—the technicians, installers, and software engineers in the field today.

A realistic solution to solving the talent pipeline gap is developing suitable, formal training and reskilling programs for people and companies that are aligned with the future technology and industry needs.

Goals and Objectives

With the goal of offering real-world learning solutions for mobility, autonomy, and intelligent transportation, The NEXT Education was created. Developed as a resource for retraining and upskilling current employees, The NEXT Education is a training firm with the mission of preparing a workforce for the mobility jobs of the future.

The NEXT Education is dedicated to equipping leaders, professionals, policymakers, and technicians with the knowledge and skills to innovate and advance global growth in these rapidly evolving fields through training that prepares individuals and organizations to compete and forge new frontiers.

The demand for knowledgeable leaders and a skilled workforce to develop and deploy those technologies is increasing. Our goal is to develop the next workforce equipped with the knowledge and skills to advance global growth within intelligent transportation, cybersecurity, and new mobility sectors. An upskilled workforce is a stronger, more competitive workforce.

Our credentials not only equip and empower individuals and organizations with the knowledge and power to develop and deploy these new technologies; they instill confidence among existing and potential partners and customers as they decide who to trust with their business.

Approach and Challenges

To ensure today's workforce is adequately prepared for tomorrow's work, training is developed to deliver the knowledge and skills employees and employers need to stay ahead of the competition through a combination of on-demand video lessons and live, online instruction. The NEXT Education courses are flexible, self-paced, current and relevant, and entertaining and are taught by instructors who work on the frontiers of the technologies they teach.

The NEXT Education has partnered with organizations working in developing and advancing intelligent transportation and new mobility systems to offer credentials, certifications, and continuing education units or professional development hour credits.

Current programs include:

- Autonomous and Connected Vehicles (CAVs), including a Connected Vehicle Professional certification.
- Intelligent Transportation System (ITS) and New Mobility Communications.
- ITS and New Mobility Policy.
- Emergency Preparedness: Transportation and COVID.
- AVs Fundamentals.
- Intelligent Transportation Fundamentals.
- Smart Cities/Communities Intermodal.

To date, there have been two obstacles the organization has had to overcome: duplicate efforts in the industry and ensuring participants are learning from the best. To ensure no effort overlap between industry organizations, The NEXT Education collaborates with others who want to share their existing expertise, research, and skills.

Today, the organization collaborates with industry stakeholders across the country, including the American Center for Mobility, Center for Automotive Research, Connected Vehicle Trade Association, IEEE, Michigan Aerospace Manufacturers Association, and more.

To ensure coursework was not only educational but was also taught by leaders in the space, The NEXT Education has partnered with leaders in the intelligent transportation and new mobility industries. These leaders give participants access to real-world insights and the ability to envision what their career paths could look like.

Results

In the four years since The NEXT Education launched its coursework, more than 900 individuals and company participants have progressed through the curriculum. The organization has worked with both public and private sector employers to upskill their current employees.

Through unique partnerships, The NEXT Education avoids reinventing the wheel by collaborating with others who want to share their existing expertise, research, and skills. In 2020, The NEXT Education partnered with Center for Automotive Research (CAR), Workforce Intelligent Network, and multiple universities across the United States—including Oakland Community College, which is known for its continuing education courses—to offer its Connected Vehicle Professional Credentialing Program to a larger audience through a three-day online course.

In 2021, the ITE and The NEXT Education announced an agreement to develop and administer a NEXT/ITE Technician and Installer Credential program. This joint certificate program focuses on intelligent transportation and new mobility for technicians and installers, working with industry leaders, planners, engineers, technicians, and stakeholders to ensure the mobility industry's pipeline is healthy, knowledgeable, and ready for the future.

About the Team

The NEXT Education was started in 2017 by Elaina Farnsworth and, although it services companies and workers across the country, is based in Metro Detroit, Michigan.

Elaina Farnsworth is the CEO of The NEXT Education, an educational company providing credentialing and training programs through a blend of video and online live instructor career development for the intelligent transportation and new mobility industries. An acclaimed speaker, published writer, and thought leader in the AV and cybersecurity industries, Elaina has been recognized as one of *Industry Era*'s top 10 Women of 2020, a 2018 Top 10 Influencer for North American Automotive Suppliers, 2015 TechWeek100: Top Tech Leaders in Detroit, and 2015 *Corp! Magazine*'s Most Valuable Professional. She serves as an advisor to the U.S. Army GVSETS Cyber Industry Chairperson and the Michigan Automotive and Defense Cyber Awareness Team. She was also appointed Director of Global Communications for the Board of Directors of the International Connected Vehicle Trade Association and is an advisor to other educational institutions and nonprofit organizations. She is passionate about the need for knowledgeable leaders, educated employees, and skilled tradespeople in the fields of CAVs, new mobility, cybersecurity, and smart cities and communities technologies and deployment. Her advocacy in these fields continues to equip others with the skills and knowledge to advance in an ever-changing world.

Mobility Hubs of the Future

Nicole Meunier, Interim Director of Development, Mercy Education Project (MEP)
Janet L. Attarian, AIA, LEED AP BD+C, Principal I Senior Mobility Strategist, SmithGroup
Amelia E Duran, Co-director, Garage Cultural
Students: Andrea Alcantar, Tania Almanza, Jennifer López Beltrán, Ximena I. Mazariegos, Nathalie Miranda, Karla Osorio, Nazli Maria Sancen Calderon, Ni'Matullah Watkins, and Nicole Meunier
Detroit, Michigan

Background

When young women in high school were asked about mobility in Southwest Detroit, they were challenged to prioritize their key concerns and consider actionable solutions. This exercise was a part of the City:One Challenge, and the Mercy Education Project (MEP) was asked to develop a design for submission. The MEP is a Detroit organization that works to provide tools for educational equity, economic stability and mobility, and cultural enrichment opportunities for women and girls who have limited access to resources. The City:One Challenge was launched in June 2019 by Ford to help tap into local resources to dream up new solutions, designs, and innovations to improve the way people get around, preparing our cities for the future. The goal of the challenge was to improve mobility in the community surrounding Michigan Central Station (see Chapter 7) so that mobility could complement existing services for transportation. The two key concerns identified by this team of students were transportation access, safe mobility in and around Detroit, and preparation for tourists and visitors. They noted Detroit's large geography of 142.9 mi^2 creates challenges of disconnection due to a lack of connected and dynamic systems in both mobility and communication, which then creates challenges for adequate resource distribution.

The youth team determined that a temporary mobility station—the MEP Mobility Hub, or THE HUB—would offer solutions to the challenges experienced in this corridor in Detroit's Corktown neighborhood.

Goals and Objectives

The team developed four goals that their mobility hub would work to achieve:

1. Increase demonstrated safety for pedestrians at THE HUB.
2. Improve connectivity between the neighborhood and downtown.
3. Increase walkability and foot traffic.
4. Increase local business foot traffic.

Through further development of these goals, the team worked to develop key performance indicators that would prove how this hub could serve as a solution for the community. They talked with local businesses near the proposed station locations to learn which days saw peak sales. The team reached out to Corktown Block Clubs to present their idea, share how it could benefit the community, and explain how THE HUB provides a safe space for everyone. THE HUB was designed to build unity between Southwest Detroit and Corktown. The team is providing an online platform to obtain residential feedback regarding what they like about THE HUB and what they would like to see at future mobility hubs and stations. The group worked with the Corktown Business Association and the Southwest Detroit Business Association to collect data on THE HUB branded coupons redeemed at local businesses along the corridor provided to visitors to encourage foot traffic in local retail.

Approach and Challenges

During the six-month pilot phase, THE HUB will provide similar services as a permanent structure, but scaled down and at a fraction of the price. The primary goals of building a temporary structure instead of a permanent station are to measure usage, foot traffic, and impact within the neighborhood, to discover the most effective permanent location, and to help establish a shared community space for residents, workers, and tourists to connect mobility in a real way.

THE HUB will provide easy, free information and access to the impact zone neighborhoods, the City of Detroit, and the various methods of mobility. THE HUB will create a safe location that will also increase access to information and mobility. Security will be provided by the Michigan Central Station security vendor as part of their daily patrol route. THE HUB will also increase economic activity in the commercial corridor of the Impact Zone through partnerships with local business associations and by creating advertisements, marketing campaigns, and other retail opportunities specially generated for the stations.

Due to the nature of development in the city, the team identified early on that the permitting process could cause a delay in project implementation. Vandalism to the structure was another challenge identified, and the team designated additional funds in their budget to make sure that the operations, security, and maintenance would keep THE HUB in optimal operation. This portion of the budget would also include adjustments or maintenance related to weather damage. The team recognizes that, due to the city's high concentration of homeless populations, there may be need for additional support to help affected Detroiters find other safe shelters.

Age was one challenge that the group experienced and could not solve. They recognized that being under 18 years old might put them at a disadvantage when executing this proposal. The MEP supported the group by including the students as active Board of Directors during the pilot and taking the direction of the young women for the integrity of the mobility station project. Because THE HUB is a brand-new concept within the city, how to classify the community space for proper permitting was a challenge. The definition of THE HUB is as follows: an installation that is a community space that functions as an extended bus station that also provides access to various methods of mobility.

Similar to stories being told worldwide, the COVID-19 pandemic was one of the largest challenges to this project. Learning how to build a community space remotely, within budget, and safe to use forced the team to think outside the box. Because of the pandemic, there was a concern that usership at the space would be low. Therefore, the team has included additional efforts for creative marketing and community outreach, as well as safety precautions, such as supplying THE HUB with personal protective equipment (PPE) and hand sanitizer (Figures 10.1 and 10.2).

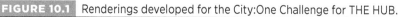

FIGURE 10.1 Renderings developed for the City:One Challenge for THE HUB.

FIGURE 10.2 Renderings developed for the City:One Challenge for THE HUB.

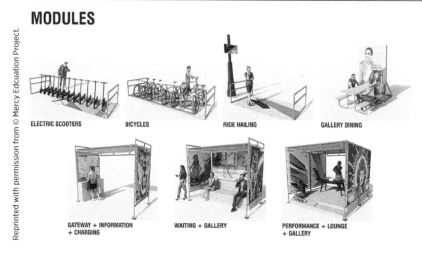

Results

The MEP's THE HUB was one of three winners for the City:One Challenge, and the project will share $250,000 in funding with AbleLink Smart Living Technologies and the Downtown Detroit Partnership. The target date to open The HUB is May 21, 2021. The winning design for these mobility stations would increase access to mobility, information, and options. Kiosks would help provide information on schedules, maps, and parking, as well as information on bikeshare and rideshare.

THE HUB was installed in the Corktown neighborhood in Summer 2021 (Figure 10.3).

Since winning the competition, all eight high school seniors have applied and have been accepted to college programs.

About the Team

Nicole Meunier, MEP Mobility Hub Project Director and Interim Director of Development and Major Gifts at MEP, has 10 years of experience building community through nonprofit development and fundraising (Figure 10.4). Meunier started her career in the arts sector in Detroit and relocated to Chicago where she worked with several nonprofits ranging in cancer research, youth homelessness, and the Chicago Theater District. Upon returning to Detroit in 2018, Meunier started working at MEP, applying the skills and resources she obtained while in Chicago. Meunier is a passionate Detroit resident and professional when it comes to doing her part in making Detroit one of the greatest U.S. cities. Meunier also serves on the Corktown Business Association Board of Directors as an executive member. She is involved in several community groups through the city and is soon to be a Certified Fundraising Executive.

FIGURE 10.3 Young women enjoying THE HUB outside of Michigan Central in Detroit, Michigan.

FIGURE 10.4 The student team from the MEP.

Janet L. Attarian, AIA, LEED AP BD+C, Principal I Senior Mobility Strategist at SmithGroup, is passionate about improving mobility in Detroit, creating equitable community development, and empowering women and youth. Playing a lead role in supporting the MEP and the eight young women who won the Ford City:One Challenge in implementing their vision for a mobility hub brings these passions together. Working closely with the young women to reflect their ideas around community, safety, and access led to a design that not only increases mobility access but supports local businesses and creates a place where people can visit and can feel comfortable, safe, and welcome. The project is innovative and forward thinking in three key ways. First, it was contemplated, designed, and inspired by the community and reflects not only what mobility means to the young women, but it empowers them to feel confident in their ability to effect change and lead. Second, creating a mobility hub that not only integrates electric charging scooters, real-time transit info, and is powered by 100% on-site solar but also provides seating, shelter, a performance stage, and murals by local artists is an amazing accomplishment for MEP, a local nonprofit helping girls thrive. Third, implementing THE HUB where the Corktown and Southwest communities come together and in front of the historic Michigan Avenue Train Station links the girls' Southwest community to downtown and reflects the potential of the many mobility and community projects planned for the area. The partnerships required to design and implement THE HUB provide a model of the integrated teams required to create, implement, and operate mobility hubs. It also displayed how to build a Detroit-specific model of how it can be done and gives the young women deep insight into how things get done and the wide range of careers available to them. The project celebrates women, from its inception by eight young women to the art that celebrates and reflects women and their community to the many women who helped make this project possible.

Because the MEP is a nonprofit, this project was able to invite large corporate partners to support the mobility hub. The project was designed by SmithGroup, built by L.S. Brinker, powered by DTE Energy, connected by Verizon Wireless, and illuminated by Laidlaw and Morgan Lighting Group. Pepperl + Fuchs Inc. provided a digital display for wayfinding and local advertisements. The project is celebrated by the City of Detroit, Ford Motor Company, Corktown Business Association, and the residents of Detroit.

About the Student Team

Nathalie Miranda studies at Western International High School and has been accepted by Wayne State University to pursue a degree in nursing.

> This project has given me the best work experience and has taught me time management, how to work with a team, and most importantly, it has allowed me to strive out of my comfort zone by trying new things. This project is important to me because I get to be a part of the change I want to make in our city.

Ni'Matullah Watkins is a student at Cass Technical High School. She is undecided on her university of choice but is interested in performing arts

> Something I learned personally from this project is that the girls and I have more in common than I realized even though we come from two different schools and areas. This project is important to me because I—being the only black female in the group—have a chance to make a difference. I have a chance to use my voice and be a part of something that other young black girls can look up to.

Andrea Alcantar is a student at Western International High School. She has been accepted to Wayne State University to pursue a career in medicine.

> From this project I learned how much public transportation has a big impact on communities. This project is important to me because I want to contribute to my community in hopes of seeing Detroit thriving again.

Tania Almanza is a kinesiology student at Albion College.

> This project is important to me because it shows my potential and allows me to give back to a community that has helped me get to where I am today. It has also shown me how to use the opportunities given to me to my advantage and to push myself forward in order to get to a better position to accomplish my goals.

Karla Osorio is a student at Western International High School and has been accepted to the University of Michigan. She is planning to study law with a focus on either political science or business management.

> What I want to do when I grow up: I'm not 100% sure what I want to do when I grow up. I plan to become a lawyer and provide a voice/help to groups and people who need to be heard. However, life might take me down a different path, but whatever happens, I know I want to help others. To me, this project was always about the community in Detroit; it is such a vast and diverse city with so much to offer, yet the people living here don't have the opportunity to witness this. The goal for this mobility hub is not only to provide a safe, efficient way to travel but also to be a paragon that connects the various in the city communities together.

Jennifer López Beltrán is a student at Albion College; her major is undeclared.

> Throughout this journey I haven't only learned to work alongside new people, I have learned to become wiser, quick thinking, relatable, and confident. It has been amazing being able to come up with a mobility hub for my community to use. Not only has this been an amazing gift to my community but to the young brilliant minds out there as an example that we can all help better our hometown with dedication.

Nazli Maria Sancen Calderon is a student at Western International High School planning to major in business and political science in college.

> Wow, how do I even begin? I truly believe this experience was something that helped shape the person who I am today. It enabled me to find my inner confidence and improve myself as a leader. Being a teen and a woman of Hispanic heritage, I have been constantly criticized and faced with stereotypes. This project taught me how to never take no for an answer and to become more self-assured with myself. I learned that by being vocal and brutally honest, it can help create change. That is why this project is very important to me because I truly believe it's making a change. With building the mobility hub, I hope it expresses how much I love and care for my community and how much I want to see it prosper. In 20 years I want to come back and see the difference the girls and I made, in addition to hopefully seeing this project inspiring others to do the same no matter who you are.

Ximena I. Mazariegos is a student at Western International High School. She has been accepted to Wayne State University to study environmental science.

> This project has given me valuable experience that I will need in a professional workplace and has taught me as long as you work hard nothing is impossible. This project is important to me because it gave me the chance to help out my community and meet new people that have helped us achieve our goal.

11

Technology Innovation

lightpoet/Shutterstock.com.

What's most interesting to me about reporting on tech and mobility right now is the collaboration that we're seeing between the "traditional" auto companies and new entrants. This collaboration is the only way forward to make the world of mobility more universally accessible, equitable, affordable, and safe.

—Alexa St. John, Transportation Reporter
Insider

Between the two of us, we are lucky enough to be able to witness some of the greatest feats in transportation innovation taking place right in front of our eyes. Our jobs give us access and front row seats to some of the most innovative technology that the industry has to offer by air, land, and sea, which are addressing some of the world's most pressing issues. This access is partially the inspiration behind this book—being able to share a glimpse into the future is the most exciting part of being a storyteller in this industry.

Technology innovation in the automotive industry is segmented into CASE (Connectivity, Autonomy, Shared mobility, and Electrification) technology, which is enabling this next generation of mobility products. According to a new study examining mobility transformation from 2020 to 2030, as predicted by the Experiences

Per Mile Advisory Council (featured in the Design and Engineering chapter), CASE technology is broken down into four focuses:

- Developing vehicles and infrastructure with connectivity that will deliver new digital experiences to consumers and leverage data in new ways to streamline operations.
- Automating mobility to reduce accidents in the short term, help improve productivity among users in the mid-term, and fully manage traffic flow in the long term.
- Creating options for shared mobility that maximizes utilization of vehicles and delivers low-cost journeys to consumers in congested areas.
- Electrifying mobility to reduce the reliance on fossil fuels and to improve air quality by replacing internal combustion engines with EVs.

It is important to note that CASE is not just specific to privately owned vehicles. This is a revolution that is taking the entire transportation industry by storm—from the electrification of bus fleets and increased mass transit platforms to reduce cities' carbon footprints to the ingenuity behind rethinking the movement of goods, along with people, that is faster and more efficient by air, land, and sea.

At its core, mobility innovation is about creating transportation options that are—as Alexa St. John put it—accessible, equitable, affordable, and safe. And we would add sustainable. These ideas have truly been at the core of this entire book. The industry is designing vehicles and shuttles that can drive themselves, allowing people mobility freedom they did not have before. There are companies focused on advancing technologies beyond our wildest dreams. And there are organizations that are dedicated to ensuring all of it is building toward a greener and more environmentally friendly future.

The case studies featured in this chapter are not meant to be an exhaustive list of the incredible innovations being developed today, but more of a sampler of various segments of mobility. From the roads to the skies and everything in between, these case studies demonstrate the ingenuity of the female minds leading this industry.

Mapping the Skies for Intelligent Air Mobility

Ana Healander, Co-founder and Vice President of Customer Success, Airspace Link

Lisa Peterson, Vice President of Business Development, Airspace Link

Background

As the world struggles with growing populations, denser city centers, and the traffic and congestion that comes along with both, the logistics around transporting people, goods, and services becomes increasingly difficult. Some believe the best way to alleviate the stress on roadways and congestion in urban areas is to rethink the spatial plane and look toward the sky.

In recent years, several large companies, including Amazon[1] and UPS,[2] have made announcements around the option to transport and deliver items via UAVs, otherwise known as delivery drones, in order to rethink avenues for last-mile delivery. These drones, which are remotely or autonomously piloted, could deliver packages, food, and other goods right to your doorstep. Not only would you be able to receive your items faster, but by bypassing the roadways and reducing congestion, drone delivery becomes a sustainable and more cost-effective option for shipping providers.

In addition to their support to the last-mile delivery industry, UAVs can be used to support visual inspections of buildings or properties, precision agriculture and livestock monitoring, construction, and monitoring of roadways and other large areas of land.

Our skyways are some of our most underused real estate for transportation. However, because of that, the skies are literally uncharted territories with little understanding in terms of governance and regulations by drone pilots and the communities they fly in.

To accomplish a large-scale commercial drone operation, there are a number of hurdles operators would have to overcome, including:

- Obtaining Federal Aviation Authority (FAA) authorization to fly in the more than 40% of the U.S. population is covered overhead by controlled airspace.

- Understanding and navigating the constantly changing temporary flight restrictions and more than 400 local government ordinances.

- Receiving approval by the FAA to fly beyond a visual line of sight by compiling safety case data, sometimes resulting in 600-page reports, which can take months to assemble.

Ultimately, businesses and operators are responsible for their own infrastructure to manage flights—a task that is daunting and nearly impossible for mass operations. Similar to the way the trucking industry relies on coordinated infrastructure efforts in the form of the highway system, drones also need a highway to operate safely, orderly, and efficiently. Especially at scale, an advanced drone operation needs a shared UAV infrastructure and participation and support from state and local governments to succeed. This is imperative for the safety of everyone on the ground and in the air.

This lack of coordinated effort in the skies makes it incredibly difficult to fly commercial drone services legally and safely and to integrate drones into both domestic and international airspace and into our communities.

In an effort to make mass UAV services possible, Airspace Link, the leading North American provider of solutions for state and local government agencies to better plan for and manage the safe integration of drones, built an unprecedented team of federal, aviation, transportation, geospatial, and GovTech innovators. This team is tasked with envisioning the way we look at the sky through advanced air mobility (AAM).

---

[1] https://www.amazon.com/Amazon-Prime-Air/b?ie=UTF8&node=8037720011
[2] https://www.ups.com/us/en/services/shipping-services/flight-forward-drones.page

Goals and Objectives

To provide drone delivery and UAV services with the infrastructure the operators need to succeed, Airspace Link set out to help communities safely manage drone integration while building new economic development opportunities for businesses and citizens. Airspace Link aimed to create a platform that simplifies the process and shortens the time it takes to receive flight authorization. To do so, the software company needed to develop a platform that would allow:

- Recreational flyers and commercial drone pilots to safely and legally fly in seconds, with authorization.
- Commercial pilots to request advanced flight authorizations by communicating their operation details directly to air traffic control, similar to manned aircraft.
- Advanced flight hazards and risk reports based on the operations, air, and ground hazards to be generated in seconds rather than following months of research.
- Pilots to connect to 726 air traffic control towers from one centralized location for real-time authorization and communication.
- Governments to hold the necessary tools to operationalize an Unmanned Aircraft System (UAS) digital infrastructure.

Airspace Link set out to become the safest and most authoritative body on drone delivery and other UAV regulations and operations.

This is a major inflection point in the infrastructure industry and the biggest disruption coming to infrastructure in 150 years. Cornelius Vanderbilt built the railroad, Thomas Edison was responsible for the electrical grid, President Dwight D. Eisenhower is credited with the modern highway, and now Airspace Link was aimed at mapping the digital infrastructure for low-altitude aviation.

Approach and Challenges

In order to simplify the process and therefore the time required for drone flight authorization, Airspace Link created two platforms for the public under a singular AirHub™ moniker. AirHub™ is the first cloud-based drone platform focused exclusively on merging the needs of state and local government with the operational planning tools drone pilots need to plan and authorize recreational or commercial operations within the community.

AirHub™ for Government provides a single authoritative data source across multiple levels of government and ultimately enables the drone industry to use the safest operational locations or routes in the community. Understanding that drone operations are already occurring within their communities, municipalities can use AirHub™ to prepare for and harness the power of UAVs. With Airspace Link's flight-planning tools, government entities can work with drone operators to publish data and communicate risks and regulations for use by pilots. It will also aggregate federal and local data into a comprehensive view of high- and low-risk areas within the community, literally mapping "highways in the skies" to better inform the local municipality and empower them to explore the additional economic development opportunities in the industry.

AirHub™ for Pilots is aimed at recreational flyers and certified replot pilots flying under Part 107, a set of rules set by the FAA for individuals offering professional drone services in the United States such as aerial photography and land surveying. Through this platform, pilots use Airspace Link's online tools to plan and submit operation details and can receive authorization in near real time. By partnering with the FAA, AirHub™ can provide authorization and safely integrate small drones into the national airspace by creating airspace awareness and providing a platform to receive automated approval to fly in controlled airspace.

The largest challenge throughout this process was gaining FAA approval, which took nine months. Now Airspace Link is an FAA-approved UAS Service Supplier of Low-Altitude Authorization and Notification Capability, providing drone pilots with access to controlled airspace at or below 400 feet.

Additionally, Airspace Link was the first company in this space to consider the ground risk in addition to the skies. They take in local authoritative data to show where hospitals, schools, helipads, and other high-risk areas are located in order to give the pilot the safest route possible. In order to provide an industry-leading standard for ground and air safety, Airspace Link partners with the American Society for Testing and Materials International, NASA, FAA, MITRE, and JARUS models.

Results

Airspace Link has seen increasing interest and growth after launching the AirHub™ platforms in April 2020. Today, Airspace Link's platform manages FAA, state, and local government drone infrastructure—providing flight authorization for drone pilots in seconds, rather than months—and is the safest, authoritative, and most trusted system on the market.

The AirHub™ platforms now service more than 300 businesses and provide commercial drone pilot authorization, paving the way with governments, manufacturers, software, unmanned aircraft system traffic management (UTM), and service providers for large-scale UAV deployments. The platform has been so successful that Airspace Link is currently operating in 17 states and multiple cities, providing government entities with a transportation planning and management tool to support drone operations locally.

In recent months, Airspace Link has begun supporting public safety operations in Ontario, California, highway road inspections throughout North Dakota, and automotive parts delivery in Michigan and Wisconsin, and is even delivering COVID-19 test kits in Syracuse, New York.

About the Team

Airspace Link consists of an expert team of GIS scientists, data scientists, geospatial developers, analysts, and accomplished web developers and developed GIS solutions—acquired by 3M in 2010—which are used in 16 states for managing transportation systems. Now with AirHub, Airspace Link will be at the forefront of ushering in a new type of mobility and transportation at low altitudes around the world.

The success of AirHub hinges in great part on two female executive leaders along with a team of three female data scientists, who are dedicated to making AAM a reality and bring a diverse and unique perspective to the drone industry.

Ana Healander is a co-founder of Airspace Link and serves as the company's Vice President of Customer Success. In this role, Healander ensures current customers are up to speed on advancements and assists with connecting customers to larger stakeholders and opportunities. Healander has 20 years of sales management and account management experience, having previously worked for Johnson Controls and Metal Supermarkets, making her well rounded in direct sales, inside sales, and customer management. She also has 13 years of experience as a co-owner and franchise manager and operator.

Lisa Peterson is Airspace Link's Vice President of Business Development and Marketing and is responsible for cultivating relationships with state and local government agencies, as well as industry solution providers who rely on Airspace Link's AirHub platform. Peterson has more than 25 years of experience in developing markets for emerging and transformative technologies in both venture-backed startup environments and large corporations such as Sprint-Nextel, Neustar, Tata Communications, and Teralytics. Prior to joining Airspace Link, Peterson was Head of Business Development for AiRXOS, a startup acquired by GE Aviation focused on intelligent avionics solutions for UAS and UAS UTM.

Autonomous Bus Pilot Deployment

Tammy Meehan Russell, Global Business Development and CAV Ecosystem Manager, 3M
Minnesota

Background

A decade ago, many connected and AV technologies were being researched and developed in the blooming CAVs Sector, and 3M was leading public and private sector conversations related to infrastructure R&D requirements for this new set of technologies to be deployed through U.S. and global teams. The 3M global headquarter is located in St. Paul, Minnesota, which is now recognized as an early adopter and developer of future-ready transportation, but that was not always the case. Many public entities in Minnesota were not informed of the realities of new CAV technology, how they could operate, or why it was important to both the quality of life of Minnesotans or economic development opportunities associated with the technology. At the time, Minnesota was a leader in the space of ITS (Intelligent Transportation System) but was far less familiar with the new world of CAVs.

In 2017, the MnDOT conducted an Autonomous Bus Pilot to prepare for automated shuttle deployments that could handle both mixed general traffic and cold weather climate conditions. The automated shuttle bus, provided by EasyMile, was the

EZ10 model. At that time, the EZ10 was equipped with high-accuracy global positioning system (GPS) and eight separate light detection and ranging (LIDAR) sensors. The LIDAR sensors included four 270-degree single-layer sensors mounted at each lower corner of the vehicle. There were two 16-layer sensors, one in the front and one in the back of the vehicle, designed to detect an obstacle in a cone-shaped zone in the front and back of the vehicle. In addition, two 180-degree roof-mounted sensors were designed to detect landmarks in the surrounding environment for localization.

Goals and Objectives

The purpose of the proposed Minnesota Autonomous Bus Pilot project was to define an AV pilot engagement plan and solicit technology partners to come to Minnesota to work with the stakeholders in safely demonstrating the technology. In order to help the State of Minnesota prepare to lead in CAV adoption, 3M worked with the state on an educational campaign with the following testing and demonstration goals.

The testing and demonstration goals included the following (Table 11.1):

1. Identify the challenges of operating AV technologies in snow/ice conditions and test potential solutions through field testing.
2. Identify the challenges and strategies of having third parties safely operate AVs on the MnDOT system.
3. Identify infrastructure gaps and solutions to safely operate AVs on the MnDOT system.
4. Prepare transit for improving mobility services through AVs.
5. Increase Minnesota's influence and visibility on advancing automated and connected vehicles.
6. Enhance partnerships between government and industry to advance automated and connected vehicles in Minnesota.
7. Provide opportunities for public demonstrations of AVs and obtain public feedback.

TABLE 11.1 Types of observations recorded during vehicle demonstration.

Testing notes	Vehicle events	Weather details
Time of Day	Sensor Activated Slow	General Observations
Date	Emergency Stop	Temperature
Person Completing Form	Intersection Stop	Feel Like Temperature
Lap Number	Manually Drove Vehicle	Wind
Testing Scenario	Battery Charging Issue	Dew Point
Start Time	Planned Start	Pressure
End Time	Planned End	Sky Conditions
Battery Temperature (in Celsius)	Planned Obstacle (i.e., vehicle, bicyclist, pedestrian, barrel, etc.)	Precipitation
Battery Charge Level	Planned Stop	Humidity
Heater On/Off	Platform Stop	Weather Source
Lights On/Off	Other Events	Visibility

Approach and Challenges

It was critical to build awareness and educate the Minnesota ecosystem on CAV technologies. This project started with bringing the winter weather autonomous shuttle project to Minnesota during the 2018 Super Bowl. MnDOT offered free shuttle rides at the Nicollet Mall between 3rd and 4th Streets from 11 a.m. to 4 p.m. on the Saturday and Sunday prior to Super Bowl LII. The automated minibuses could hold up to 12 passengers, but these rides were limited to six seats. Travel speeds floated about 5 mph along preprogrammed GPS-guided paths. The top speed of these minibuses was 25 mph, which was not reached during the pilot.

Though much research had been conducted up to this point, there were not extensive findings on how AVs would perform in winter weather conditions. This posed several unique challenges for the pilot. In order to conduct a comprehensive pilot, it was split into three phases:

1. MnROAD Track Testing: This phase provided a controlled environment in which to test the automated shuttle in a variety of winter weather conditions. The track was a 2.5-mile low-volume closed loop at MnROAD Test Track.
2. Downtown Minneapolis Demonstration: This phase allowed the key stakeholders and public to ride the automated shuttle and give feedback on their experience.
3. Additional Demonstrations: This phase allowed a wider variety of stakeholders to ride the automated shuttle and demonstrate its capabilities in a variety of environments including demonstrations at the 3M campus and the State Capitol.

Results

The successful demonstration of a CAV shuttle in various Minnesota locations in 2017–2018 included more than 3,300 riders. Of those 3,300 riders, 238 participated in the stakeholder tours at the MnROAD[3] test track in December of 2017. A total of 1,279 riders participated in the three-day public event in January 2018 (the week of Super Bowl LII). A total of 216 riders participated in riding the automated shuttle bus at the State Capitol Demonstration. Four hundred seventy-nine riders participated at the 3M campus: 267 riders in the City of Rochester, 413 in Minneapolis, and 454 at the University of Minnesota.

Minnesota now is in the second phase of the appointed Governor's Advisory Council for CAV, and, in 2020, formally rolled out the Minnesota CAV Innovation Alliance with five new subcommittees.

Below are the shuttle's performance-based findings from the 2017–2018 pilot:

1. Clear weather/bare pavement: The automated shuttle bus performed well in periods of clear weather and bare pavement. The bus interacted well with other cars, pedestrians, bicycles, and obstructions. Some sensor slowdowns occurred due to blowing dust, weeds, and snow from the shoulder.

---

[3] MnROAD is a pavement test track made up of various research materials and pavements owned and operated by the Minnesota Department of Transportation, working with its partners.

TABLE 11.2 Scenarios and findings from vehicle interactions with pedestrians.

Scenario	Findings
Bicycle ahead at varying speeds traveling in the shoulder, near edge line, in center of lane, or near centerline	Good interaction. Bicycle detected, automated shuttle bus did controlled slowdowns as needed
Bicycle traveling in the same direction in the parallel lane adjacent to the automated shuttle bus or passing or being passed	Good interaction. Performed slowdowns or stops when necessary
Bicycle traveling in the opposite direction in the opposing lane at varying distances from the centerline	Good interaction. Performed slowdowns if an opposing bicycle was detected too close to the automated shuttle bus
Bicycle crossing the travel lane in front of the automated shuttle bus	Bicycle detected. Automated shuttle bus did a controlled slowdown.
Bicycle crossing travel lane in front of automated shuttle bus and stops in the center of the travel lane	Bicycle detected. Automated shuttle did a controlled slowdown and stopped. Stop distance measurement from bumper to bicycle foot pedal = 6.5 feet

2. Light snow: Performance was similar to performance on bare pavement, and the slowdown was similar to blowing snow and snow kicked up from tires.
3. More severe snow conditions: Falling or blowing snow was detected as obstructions to the sensors, causing slowdowns in order to avoid perceived collisions.
4. Rain/fog: Mild wet conditions did not appear to impact the vehicle's performance.
5. Controlled snowmaking conditions: Two snowmaking systems were used for testing in both mild and below freezing conditions. During this controlled experiment, the bus performed sensor-activated slowdowns when trying to navigate the man-made falling snow but was able to recover its automated function and proceed on the route when it stopped "snowing."
6. Varying pavement conditions: Snow conditions led to wheel slippage, mostly at higher speeds. The performance was not impacted by varying lighting conditions.
7. Interaction with obstructions and other vehicles: See Table 11.2.
8. Interactions with pedestrians.
9. Interaction with bicycles.

Battery performance was also measured. Measurements were taken at multiple points during the demonstration to understand how winter weather would have affected the battery life. The colder weather did prove to discharge the battery faster, especially in subzero temperatures when the vehicle heater was in use.

About the Team

Tammy Meehan Russell is President and Chief Catalyst of the PLUM Catalyst (Figure 11.1). Through her role as both Founder and Chief Catalyst of the PLUM

FIGURE 11.1 Tammy Meehan Russell poses with one of the CAV shuttles.

Catalyst, she understands that building awareness and engagement with the CAV technology is still a high priority. In 2019, recognizing the need to help other mobility technology companies navigate the roads to government pilot deployments, Russell founded the PLUM Catalyst. Through her work with the PLUM Catalyst, Russell continues to work on CAV awareness and education through a partnership with MnDOT, helping to build and support the Minnesota CAV Innovation Alliance, supporting upcoming CAV-related pilot and deployment projects, and promoting transportation equity in CAV technology through the support of the new Minnesota G.E.T. (Gender Equity Transportation) Collaborative.

Russell is helping grow the Minnesota ecosystem of public-private partnerships through her work with the University of Minnesota's Center for Transportation Studies CAV Strategic Research Planning. She has been successful in bringing private technology clients to Minnesota through the help of the MnDOT CAV Challenge program and helps prepare the industry's future workforce through partnerships with local career pathways programs and the state's Department of Employment and Economic Development using innovative programs like the Minnesota Job Skills Partnership.

Working with Russell, the University of Minnesota's Center for Transportation Studies has created their new CAV strategic research plan and have approved additional infrastructure funding for CAV-related research investments as well as helping to build a strong university partnership with the local CAV technology and applied research partner, VSI Labs.

Russell also continues to expand the breadth and depth of bringing this technology to those who need it most by partnering with other industry-leading AV providers like May Mobility and introducing the future mobility technologies through pilot accessibility-focused projects and career pathways programs in rural communities in Greater Minnesota.

Russell holds a bachelor's degree from the University of Minnesota and a master's degree from the University of St. Thomas. She volunteers on the Minnesota Council and Speakers Bureau for Feed My Starving Children and serves as a board member for Crossroads Church. She lives in St. Paul, Minnesota, with her supportive husband, four amazing kids, and adorable new puppy.

The PLUM Catalyst is a strategy and technical market development boutique disadvantaged business enterprise consulting firm operating at the intersection of the future mobility technology companies, academic and research entities, and governments who want to prepare for and deploy the latest in mobility technologies. Russell founded the PLUM Catalyst to be a catalyst for change in the new mobility and smart community initiatives. Prior to starting the PLUM Catalyst, Tammy founded the 3M Connected Roads program, where she navigated her team through various global public-private partnerships to help with developing, testing, and demonstrating new roadway infrastructure technologies specifically designed for the future mobility solutions. While at 3M, Russell also started and led the 3M CAV Sensing team, which focused on providing materials to optimize and protect CAV sensors in various scenarios and weather conditions. She championed 3M's support of the Minnesota winter weather testing project for the 2017 Minnesota autonomous shuttle project and was a strategic partner with various global CAV corridors and test tracks. She has led and inspired her technical development teams in the filing of more than 100 patent applications related to future mobility solutions. Under the PLUM Catalyst, she partners with technology, government, research, academia, and nonprofit entities to help design and bring mobility solutions for those who need it. The PLUM Catalyst also contributes to multiple state and national task force efforts, including the Texas CAV Task Force: Licensing, Data, and Education Committees and the new Minnesota CAV Innovation Alliance: Workforce, Data, and Education committees.

Fun fact: Russell's background on video calls is real: Tammy's husband, Mike, painted custom PLUM Catalyst mobility cityscape paintings for the accent wall in her office!

Additional Commentary

This submission is a part of the G.E.T. Collaborative, a group based in Minnesota that is focused on identifying, funding, and promoting research related to gender and equity in transportation. The G.E.T. Collaborative started with a spark of curiosity when Ania McDonnell, Policy Analyst at Flaherty and Hood, developed an interest in gender issues in transportation as a student at the University of Minnesota's Humphrey School of Public Affairs. This spark resulted in papers and articles published by the school's Center on Women and Gender in Public Policy and culminated in her 2020 master's thesis, "Gender Mainstreaming City Comprehensive Plans," with a transportation focus. Following the successful publication of her thesis at the University of Minnesota, McDonnell's work has gained significant attention in transportation planning and policymaking circles, creating momentum behind a vision to improve gender considerations in transportation planning. McDonnell and her thesis advisor Frank Douma reached out to their networks to bring together more stakeholders to meet and discuss other ways to work on this topic. Within less than a year, members have collaborated to successfully create research proposals and gain funding. Building on this energy, McDonnell, Douma, Russell, and the other founding members continue to spread the message of gender and transportation research through presentations to groups interested in the topic.

The demonstration of the 2017 Minnesota Automated Shuttle Bus was conducted by EasyMile, the vendor chosen by MnDOT, and oversight of the demonstration was performed by WSB and AECOM staff. The Minnesota Autonomous Bus Deployment and the Minnesota CAV ecosystem growth and support could not have been successful without the entire team of visionary leaders and champions—especially noting Kristin White, Janelle Borgen, Laurie McGinnis, Susan Mulvihill, and Myrna Peterson. Tammy's leadership with 3M in CAV and future mobility discussions would not have been possible without support from her mentors and colleagues such as Gayle Schueller, Cristina Thomas, Ellen White, Joline Bogdan, Meredith Moore Crosby, Pam Henderson, and many others. *See Acknowledgments for the complete list.*

Autonomous Trucking, 2018

Anusha Kukreja, Consultant for Future of Mobility Practice, Deloitte
Seattle, Washington

Background

Through Deloitte's professional services work, an opportunity was identified regarding the autonomous movement of goods, specifically in trucking. The majority of efforts had gone toward building the technology with fewer resources dedicated to integrating into the often-fragmented supply chain operations of fleets and shippers. Deloitte's leadership in the supply chain and mobility space positioned the firm to take on the challenge of bringing the autonomous trucking ecosystem together to prepare for the integration and scaling of this new technology.

Even as a junior resource, Anusha Kukreja was empowered by her leadership to drive her team's efforts and work in collaboration with ecosystem players to develop the opportunity. The work also allowed Kukreja to leverage her prior exposure to the challenges faced by global supply chains. She kept asking, "Say the tech works, what else needs to be in place?"

Goals and Objectives

The initial phase involved research and the formation of hypotheses. Kukreja conducted the initial market research and identified a set of challenges she believed the firm could help with. For example, Kukreja had previously been staffed on a project to advise a client on locations for their future distribution centers. The work involved extensive modeling, and capital investments on this scale were budgeted years in advance. Since autonomous trucks were expected to have an increased daily driving range due to a number of factors, Kukreja realized that the optimal locations for distribution centers could be completely different for clients using the new technology. This factor was not yet being accounted for.

Once the narrative around this and similar challenges had gained traction with the leadership at Deloitte, the goal became to work with stakeholders across the value chain on the identified opportunities. Specifically, the team aimed to answer:

- If the technology works, what else needs to work for this innovation to scale?
- How can we help the necessary stakeholders understand the implications of autonomous vehicles for them?
- How can we convene the ecosystem to work together on piloting and implementing solutions?

Approach and Challenges

After conducting the initial research and creating a pitch deck to share with the internal team, Kukreja led the planning and execution of an event to further validate the hypotheses. Deloitte cohosted the Autonomous and Integrated Freight Forum, which involved approximately 75 executives from more than 40 organizations representing the full stakeholder spectrum for goods movement coming together to learn and discuss the opportunities of integrating these new technologies.

After completing this milestone, three more full-time staff members were brought on board to support the project. With the strong support of her leadership, Kukreja was able to take on a manager-level role in the team and grow her business development abilities.

The team took the following approach:

- Learn by doing: By conducting a project that included a player in the conventional trucking ecosystem (nonfleet) and various autonomous trucking startups, the team developed a new perspective around what autonomous trucks would need in the future and how conventional players could pivot to fill in some of those gaps. This project was set to launch in spring 2020 but was pivoted to be conducted completely virtually due to the pandemic.

- Grow the perspective: Through ongoing stakeholder engagement that ranged from insurance providers to small and large fleets, the team identified the impacts of autonomous trucking on current business plans. They created roadmaps for how those challenges and impacts could be addressed. The most critical aspect of this phase was educating internal stakeholders and external clients that the technology impact needed to be considered today even before the technology became available.

- Startup dialogue: Startup engagement was another critical aspect of the discovery and solutioning process. Through those conversations, the team gained a better understanding of how the tech was progressing and the challenges to scaling it as a business.

While unique for a consulting firm, the entrepreneurial and collaborative spirit of the team forged the necessary ecosystem to further enhance the concept.

Results

After a year of research, stakeholder engagement, and strategic planning, the foundation has been solidified. The firm is continuing to grow the team and is conducting research on the future of autonomous trucks to prepare businesses today. Before leaving Deloitte to pursue her MBA at INSEAD, Kukreja coauthored a white paper on the perspective her team had developed, Driverless Autonomous Trucks Lead the Way, and became known as an expert on the topic.

At the industry level, this work has accelerated thinking in the freight and shipping industry for both startups and incumbent stakeholders on the implications, challenges, and opportunities autonomous freight can offer.

About the Team

Anusha Kukreja is currently an MBA candidate and Forte fellow at INSEAD in Fontainebleau France. Prior to graduate school, she spent five years as a strategy consultant at Deloitte, primarily in the Supply Chain and Future of Mobility practices. Most recently, she was responsible for launching and growing Deloitte's autonomous truck offering. Anusha is passionate about innovation ecosystems, transportation tech, and mentoring. She graduated from Carnegie Mellon in December 2015 with a B.S. in Economics and Statistics.

Deloitte is a global professional services firm specializing in audits, consulting, financial advising, risk advisory, tax, and legal, and is home to an internal think tank called GovLab.

Additional Commentary

Historically, supply chain—especially trucking—is a male-heavy industry. For the first six to eight months, I was usually the only female on my calls. I was also always the most junior person present. But with the support of my incredible leadership, I was able to be a credible voice in the discussion, no matter how senior the group. I have made an effort to pull in more women at Deloitte and am happy to say that I am no longer the only female on most of my calls. That is one of my biggest accomplishments, personally.

All efforts in consulting are team based, so I would be remiss in not mentioning that I have had a fantastic set of leaders and teammates who have empowered me and given me the runway to grow this work from an idea into a new focus area for our Future of Mobility practice. It is not often that a junior staff member gets to take the lead as much as I was able to, and this work has reinforced my interest in focusing my career in the mobility space.

12

Pivoting for the Pandemic

lev radin/Shutterstock.com.

My CEO once shared that his six-year-old described what we do as "we fix the world when it's broken." The pandemic shined a bright light on what was broken—access to broadband, access to quality goods and services, access to gainful employment and many, many other limitations. And the toughest of all, how all these limitations disproportionately impacted underserved communities and the inequities therein. What I know from experience is that it is at our lowest point that we are most innovative. I look forward to the future because I know we can, and we will rise to the occasion.

—Denise Turner Roth, President, U.S. Advisory Services
WSP USA

When we first started writing this book in early 2020, we lived in what now feels like a different world—a world where we spent time together in the same room, a world where we did not think twice about ridesharing or ride hailing with friends or shaking hands with a stranger, a world without the novel COVID-19 virus. So much has changed since then.

Today we know all about the havoc the pandemic has wreaked on families, communities, businesses, and economies around the world. From the loss of jobs to the loss of loved ones, people have endured immense heartache and grief during this time. At the time of writing this, we have amassed more than 3 million deaths globally associated with the pandemic, with more than 500,000 of those deaths[1] in the United States, and it is anticipated that that number will continue to grow in the immediate future as well as for years to come. Just as we could not have possibly predicted the lasting ramifications COVID-19 would have on our society, we also could not have predicted just how important transportation would become in combating the deadly virus and keeping our communities safe.

As we continued to work through this book and virtually meet with women across the country, we realized we would be remiss not to raise special attention to the swift action that was taken by organizations, communities, companies, and municipalities. Almost overnight, people across the country sprang into action in order to provide medical attention and safe transportation options to medical providers and testing facilities, as well as to distribute life-saving medical supplies and vaccines. Many of the automakers and auto suppliers put their manufacturing expertise to work by pausing production of vehicles and instead focused on supporting Americans' most vital needs.

- GM, under the direction of CEO Mary Barra, stepped up to manufacture more than 30,000 ventilators for use in U.S. hospitals and produced more than 10 million face masks and latex-free face shields, protective gowns, and aerosol boxes to address the severe shortage of PPE in the health-care system.[2]
- Ford Motor Company also did its part to address the country's needs with the production of 120 million adult- and child-sized face masks, 50,000 ventilators, 32,000 air-purifying respirators, and 1.6 million washable isolation gowns.[3] The automaker also funded a public service announcement that combats the spread of misinformation about the vaccine among multicultural populations.

As we wrap up this manuscript in the spring of 2021, almost a year to the day in which our home state of Michigan went into its first "stay at home" order, we are still actively in crisis mode. Although the vaccines are currently being distributed, there are still people getting unnecessarily sick and dying daily. Today's health-care workers are still challenged for resources, Americans lack access to medical services, and vaccine distribution is proving to be difficult in some areas—especially rural and low-income communities. If there is one thing that has been cemented in our minds as this past year has unfolded, it is this: a public health crisis is a transportation crisis.

The following case studies demonstrate how mobility and transportation have remained a vital tool in combating the virus, caring for those in need, and getting the much-needed medical supplies and vaccine doses across the country and into communities.

---

[1] https://www.nytimes.com/interactive/2020/us/coronavirus-us-cases.html
[2] https://www.gm.com/coronavirus.html
[3] https://media.ford.com/content/fordmedia/fna/us/en/news/2021/04/07/ford-fund-launches-covid-vaccine-psa.html

Transportation Assistance for Tribal Communities in Remote Communities during COVID-19

Nancy Scheinman-Wheeler, Past President, Rotary Club of Towsontowne, Baltimore, Maryland
New Mexico, Utah, and Arizona

Background

Covering more than 17 million acres across New Mexico, Utah, and Arizona, the Navajo Nation and White Mountain Apache tribes account for nearly 215,000 of the regional population. This population differs culturally and socioeconomically from the surrounding communities. Within the White Mountain Apache community, households are more than three times more likely to be headed by a female member of the family and about 22% are multigenerational households, meaning several generations are living under one roof. In the Navajo Nation, poverty rates are about three times the national level. Thirty-five percent of those children living in poverty are at an increased risk of death than anywhere else in North America.

In forgotten corners of our country, Native Americans live in abject poverty. Multigenerational families live together in substandard housing, in remote communities often linked with unpaved and unmaintained roads—with little to no mass transit or public transit options available.

In 2019, the COVID-19 pandemic struck the United States, quickly affecting the entire country, from affluent communities to even the most remote. In rural tribal communities in the Southwest, Native Americans were not only experiencing health emergencies but also fighting for cultural survival. Elders in the native community were dying from COVID-19 at an alarming rate and exponentially increasing the loss of cultural knowledge that is typically passed down by elders. Native Americans have suffered the greatest losses of any group in the country.

To bring aid to the Navajo Nation and White Mountain Apache tribes during this difficult time, the Rotary Club of Towsontowne devised a plan guided by Rotary's motto: Service Above Self. Rotary International is an organization with a global network of more than 1.2 million members who aim to solve problems and create lasting change around the world, especially around issues including promoting peace; fighting disease; providing clean water, sanitation, and hygiene; maternal and child health; supporting education; and growing local economies.

Goals and Objectives

At the time of publication, the Rotary Club of Towsontowne was in the process of applying for a Rotary International Global Grant to provide reliable transportation for COVID-related services and build capacity for Native American communities in remote areas to take the lead in their own public health. Through the grant, Rotary clubs partnering together hoped to bring relief during the pandemic, which had taken a disproportionate toll on the remote tribal communities, devastating tribal areas with disease rates reported much higher than the U.S. average.

This unique initiative to fund technology and vehicles is unlike anything seen in existing research. Most foundations and federal grants do not fund the purchase of equipment costing more than $5,000, and never has one agreed to support purchasing vehicles, which the Rotary Club of Towsontowne and its partner clubs are doing for the Johns Hopkins Center for American Indian Health's Native American service providers. Often individual donors are more interested in funding people and programs, whereas the Rotary's unique interest in funding "things" and not "buildings and positions" was a perfect match.

Approach and Challenges

Through the grant, Rotarians sought to provide multiple passenger vans and pickup trucks, 10 tablets, and two MiFis (devices that provide mobile wireless internet) for five identified sites including the White Mountain Apache Reservation and four Navajo Nation communities including Shiprock, Chinle, Fort Defiance, and Gallup—each with a population of approximately 17,000 to 25,000 people.

With the vehicles and equipment, the tribal communities and medical staff would have the transportation means needed to access COVID-19 testing sites and vaccination centers, as well as transportation for future medical and health-care needs post-pandemic.

During the planning stages, the Rotary was competing against many other pandemic-related fundraising efforts, and although none were addressing the transportation needs of health-care providers for the tribal communities, it still presented a challenge.

Next Steps

At the time of publication, the Rotary Club of Towsontowne was in the funding phase of the project but looks forward to reaching their goal thanks to social media, a GoFundMe campaign, and the support of Rotary Clubs across the globe.

Additionally, Rotarians recognize that the effects of COVID-19 will linger in these remote communities long after all residents have been immunized. The purchase of reliable vehicles will allow Native American service providers to mobilize to provide home visits to young mothers and families and continue to deliver supplies and health services, including mental health services, for many years to come.

About the Team

Coordination of this project was led by Nancy Scheinman-Wheeler, a member of the Rotary Club of Towsontowne in Baltimore, Maryland, after she moved to Sante Fe, New Mexico, and proposed the grant project based on her experience within a local tribal community.

Nancy Scheinman-Wheeler, an accomplished artist, had previously worked on an art project over a period of three years with tribal elders, administrators, and high school students in the remote community of the Alamo Navajo Indian Reservation in Magdalena, New Mexico, which is where she first started gaining a better understanding of tribal communities in the Southwest. The project involved transforming the walls of the Alamo Navajo Community School with colorful, hand-painted murals depicting the important landmarks of the tribe and dioramas of the Navajo Creation Story, as well as a Code Talkers Library and a trading post. She was able to learn and witness firsthand the struggles of the Navajo people.

Scheinman-Wheeler had served as President and Foundation Chair of the Rotary Club of Towsontowne in Baltimore, Maryland. She created the club's See2Learn Project, an educational literacy and vision project that partnered with the YMCA of Central Maryland, and she also created a project supporting a shelter for homeless women in downtown Baltimore. She played an integral part in her Rotary Club's National Program to Alleviate Corneal Blindness in Bangladesh, a $500,000 global grant, and initiated a Rotary friendship project in Kathmandu, Nepal. For more than 20 years, Scheinman-Wheeler's business, NS Studios, an experiential learning environment design firm, created transformational school interiors in Maryland; Washington, DC; Pennsylvania; and New Mexico. She taught at the Maryland Institute College of Art, most recently in the Master in Professional Studies in the Business of Art and Design Program. Scheinman-Wheeler has exhibited her artwork in Santa Fe and in galleries and museums throughout the world.

Pivoting a Mobility Venture Incubator to Serve the Community

Kristin Welch, Strategy and Operations, Ford X

Background

Born within Ford Motor Company in 2018, Ford X is the automaker's venture incubator for mobility startups. It brings together entrepreneurs, designers, and engineers to envision, build, and validate new mobility ventures and business models to shape the future of transportation. To date, Ford X has graduated several early-stage projects from the program including a dockless electric scooter sharing company and a digital shopping experience for purchasing used Ford vehicles.

As the pandemic worsened globally and the region surrounding Ford's headquarters was dealing with an overwhelming outbreak, Ford X opted to pause their existing ventures to focus on what communities across the United States needed most: COVID-19 resources and response initiatives.

Goals and Objectives

Initially, the primary goal was to find the most immediate opportunity to assist in the health-care space during the pandemic. When Ford was approached by Wayne State University, Wayne Health, and the Arab Community Center for Economic and Social Services to assist in mobilizing COVID-19 testing they were already providing at stationary sites, Ford X got involved. This was exactly the opportunity they were looking for to make an impact.

They were able to mobilize testing services in an effort to help the program partners ramp up testing capabilities, especially at the beginning of the pandemic when testing site locations were overwhelmed. The team put their attention on communities of color, focusing on increasing trust in the testing process, bringing testing to the people, and informing individuals of their results. This was incredibly important as the virus was disproportionately affecting people of color.

Ultimately, Ford X was operating under the concept that they could help fundamentally change the nature of health-care provisioning for the better. By developing vehicles to mobilize health care and rethinking the model of medical services in tandem with health-care providers, Ford X was able to bring medical services to people as opposed to moving people to a central location. This model would prove to be a more effective and affordable way to improve population health during the pandemic (and beyond), especially in underserved urban communities and in rural areas where hospitals were far apart. By coming into the community, they could address access as well as increase trust in the system.

Approach and Challenges

The COVID-19 pandemic highlighted the racial and socioeconomic disparities in health care that previously existed, including the lack of access to hospitals, lack of insurance coverage, and mistrust in the system—all of which affected the health and safety of the community.

Funding allocated through Ford X allowed a 7–10-person team the time and resources to create mobile testing sites using customized Ford Transit vans—originally with Lincoln Personal Driver vehicles—to move clinicians and the service to the mobile testing sites. This clinical customization includes refrigeration, Wi-Fi, lighting, storage, work stations, increased power supply, and more.

Throughout the process, the Ford X team and program partners reengineered the vehicles in a way that would allow the clinical staff and their service offerings to move efficiently and safely. They designed and built numerous versions of the vans and placed them in the field for immediate testing and feedback from the clinicians and users and adopted changes as needed to ensure the best approach was being used to serve those in need.

To complete the vehicle build-outs, Ford X worked with the right development and design partners, including three different upfitters, and are now looking into a national network of Qualified Vehicle Modifiers and dealerships. These partnerships have helped Ford X to quickly build out vehicles for immediate COVID-19 testing and also to build based on findings and designs and to repurpose the vehicles for vaccine distribution and the future of mobile health care.

The access problems in Detroit were just the tip of the iceberg for what other communities in other states were dealing with. As the pandemic progressed, this program was adapted to fit other communities, and Ford X was able to bring their mobility expertise to work at the national level.

Later iterations of this program serviced communities in Atlanta, where Ford X partnered with CORE (Community Organized Relief Effort), a community-organized relief effort led by U.S. actor Sean Penn, also a funder of the Wayne Health–led efforts in Michigan. With the partnership with Ford X, CORE worked to mobilize a network of doctors, emergency workers, and government officials to take immediate action in the fight against the COVID-19 pandemic.

Results and Next Steps

Partnering with the correct organizations was key. Today Ford X has worked alongside dozens of key corporations, governments, and hospital systems to identify communities where help was most needed. In total, the team mobilized clinicians and mobile testing sites serving more than 35,000 people in Michigan alone.

Additionally, the Ford X team is looking toward other ways they can impact and change the way health care is provided in the future, including unique business models to improve access to mobile health units and working to ensure commercial customers' needs are met and that learnings from the mobilization of testing and vaccines during a pandemic are used for the betterment of health-care provisioning in the future.

About the Team

The multidisciplinary Ford X team working on this project throughout the COVID-19 pandemic was composed of engineers, data scientists, web developers, design thinking experts, and user experience and on mobility projects with societal impact.

Kristin Welch is a veteran of the mobility ecosystem both in southeast Michigan and beyond. For 2 1/2 years, she helped lead the ride-sharing startup SPLT, which was part of the Techstars Mobility 2015 cohort, through growth, globalization, and acquisition by Robert Bosch GmbH in February of 2018. Subsequently, she has been an advisor and strategist for numerous startups, including the autonomous delivery vehicle company Bedestrian. Welch coaches founders through the New Enterprise Forum, is a Techstars Mobility All-Star Mentor and a venture investor, and judges multiple pitch competitions nationwide. Prior to working in the startup ecosystem, Welch worked at Deloitte in global leadership roles in Boston, Zurich, London, and New York City. In 2015, she obtained an Executive MBA from the Ross School of Business.

City of Detroit's COVID-19 Transportation Response

Stacey Matlen, Senior Mobility Strategist at WSP and the City of Detroit Office of Mobility Innovation (OMI)
Hind Ourahou, Senior Mobility Strategist at WSP and the City of Detroit Office of Mobility Innovation (OMI)
Detroit, Michigan

Background

By the end of March 2020, Detroit was an early hotspot for COVID-19, with 27% of Michigan's total cases of COVID-19, and only 7% of the state's population. One year later, Detroit has had nearly 692,000 confirmed cases and more than 16,000 COVID-related deaths. In addition to immediate health costs and the significant losses from the community, the pandemic also has cost the City of Detroit financially nearly $95 million. City departments worked aggressively to make sure that care and responses were appropriate and comprehensive in order to serve and protect as many Detroiters as possible.

The City of Detroit is 143 square miles and has a population of nearly 675,000. Of this population, approximately 25% depend on transit services or other means (do not have access to a vehicle) to get around on a daily basis. The DDOT and the City of Detroit took the risks of COVID-19 very seriously, and similar to many other transit agencies across the country, DDOT reduced bus service to protect both operators and riders. At the same time, the Q-Line and the Detroit People Mover, two small transit networks serving Downtown and Midtown Detroit, halted service. While this was a safety solution, it added more barriers to a transit network that was already looking at ways to improve before the pandemic. The City of Detroit's Office of Mobility Innovation (OMI) understood the need to provide transportation assistance to Detroiters, but especially to the essential workers who depend on DDOT to get to and from work.

Goals and Objectives

The goal of the program was to help essential workers who had transportation access disrupted as a result of the COVID-19 outbreak. Outreach was conducted to better understand perceptions and reasons why participating essential workers would or would not be interested in using new mobility services to support their commuting needs. The outreach helped identify themes and patterns in user feedback to contextualize and understand user experience.

Approach and Challenges

In order to understand what transportation services were needed for Detroit's essential employees, the OMI needed a way to communicate with essential employees to understand their commuting patterns and transportation needs. To do this, the OMI created a streamlined transportation enrollment form that was initially sent to three employers with large residential, transit-dependent essential employees—two large hospital systems and one large grocery store: the Detroit Medical Center, Henry Ford Health System, and Meijer. The program has since expanded to other employers.

Print and web materials were created with detailed information about enrollment options. Human resources departments distributed these assets through employee email marketing campaigns and additionally posted flyers in community spaces such as break rooms. The program also was advertised through phone calls and text messages.

As employees enrolled, the OMI used information such as shift time, pickup location, current commuting mode, and desired commuting mode to inform the creation of new pilot programs to respond to those needs. Using this information, the OMI was able to immediately respond to essential employees who commuted at night by augmenting an existing pilot with Lyft and Detroit Cab Company to provide ride credits to help with night commutes.

Concurrently, the OMI used this enrollment form to create two new pilot programs to respond to needs. The first was an e-bike and e-scooter leasing program for employees with short commutes under six miles. The second was a shuttle service for day and night services. The shuttle service offered $2 rides for employees commuting at night and had a ride home, as well as for daytime riders who could not use the DDOT bus service due to lower capacity limits. By using the single enrollment form, the OMI was able to provide a menu of transportation solutions to these essential works who were still working.

Results

There were 63 participants in the e-bike and e-scooter leasing program, 59 on e-bike and 4 on scooters. One hundred sixty individuals were reached through phone calls, and 406 surveys were collected in total.

For the Via shuttle service pilot, from July to December 2020, 4,462 rides were provided, 129 of which were provided by the Via services.

Multiple rounds of qualitative and quantitative feedback were collected from both pilots. Of 200+ phone calls conducted from July to August 2020 to either inform or follow up about COVID-19 essential employee transportation pilots, the city documented a few key lessons learned about why people did or did not want to use the service:

- A few items that were initially considered barriers included cashless payment and data plan access. While both the micromobility pilot and the Via required a credit/debit/prepaid debit card to use, no one reported issues with the payment platform. Additionally, smartphone access was encouraged but not required for both programs; no one reported smartphone/data access as a barrier to use.

- However, through phone calls, the OMI discovered two key barriers:
    1. COVID-19 safety concerns: When asking people because they had not yet tried the Via service, many mentioned that they were hesitant to share a ride with another person. However, once safety and cleaning protocols were highlighted (50% capacity in vehicles, masks required, vehicles cleaned between uses), employees reported they were much more likely to want to try the service.
    2. Hesitancy to try e-biking or scootering as a commuting method: Many employees had not considered biking or scootering as a valid commute form to work and had many questions about how an e-bike works and how they could use it to commute. After explaining the fundamentals of biking and scootering, the OMI saw an uptick in enrollments.

Overall, it took personal phone calls to explain programs and talk through concerns to gain trust and share the value of the programs. While these phone calls could be time intensive, they were well worth the effort, as this was the main way we were able to enroll employees in the program.

Due to the ongoing nature of the pandemic and the continuation of services, final results are not available at this time.

About the Team

Stacey Matlen is Senior Mobility Strategist at WSP and the City of Detroit's OMI. With a public health and startup background, Matlen is passionate about increasing access to health-enabling activities through improved mobility solutions. Matlen is a Senior Mobility Strategist for the City of Detroit's OMI where she works to understand Detroit residents' core mobility challenges and develops and launches sustainable new mobility solutions to address these needs. Her approach is centered on improving access to mobility options while concurrently assessing economic viability and enabling the growth of the mobility industry in the city through public-private partnerships and entrepreneurial enablement.

Hind Ourahou is Senior Mobility Strategist at WSP and the City of Detroit OMI. With 10 years of design thinking practice, Ourahou brings a different approach to her work as Senior Mobility Strategist for the City of Detroit. Ourahou's international experiences in economic development and entrepreneurship proved fundamental to shaping how she leverages mobility innovations. Ourahou strongly believes that designing the mobility experience anywhere is essentially an exercise in empathy in which the thorough examination of behaviors and spaces allows us to see beyond binary answers to our mobility challenges.

Afterword

So Where Do We Go from Here?

Jessica Robinson
Co-founder and Partner, Assembly Ventures
Co-founder, Detroit Mobility Lab and Michigan Mobility Institute

I often start presentations on the subject of mobility with a slide of three photos. In the first, a young woman in her twenties walks down the street listening to music on her earbuds. The sun is shining as she looks up and out of the frame smiling and eyes twinkling. She is bundled up in a heavy coat, but you can just tell she is loving life as she is on the move. In the second image, another woman stands up in the back of a parked convertible. Sunglasses perched on the top of her head and eyes closed, she faces the rear with arms spread wide as if she is soaring. In the final image, an open road with rolling green hills on each side suggests escape and adventure.

This is the true promise of mobility: freedom.

Yet as anyone who sits in traffic wonders whether the bus came early, lands at the end of the parent pickup line at school, waits in a long line of trucks at the weigh station, peers out the window hoping that their groceries will be delivered soon, takes time off from work to help a family member get to a doctor's appointment, or hurries to fill their gas tank while rushing to a meeting knows, *mobility often is anything but freedom.*

While often featured in ads and images like the ones I just described, women have not always had a seat at the table when key decisions in mobility are being made. Sometimes we are complete afterthoughts—and with deadly consequences. Automobile crashes are the single largest cause of death for pregnant women.[1] Even so, vehicle crash tests generally use dummies based on the "norm" of a medium-sized male body. The automotive industry has been using crash test dummies since 1971,[2] but the first pregnant female crash test dummy, known as MAMA 2B, was developed only in 1996. As recently as 2004, researchers at the University of Michigan Transportation Research Institute began testing a new retrofit[3] to better understand crash impacts on not just a mother but also her fetus. *Inclusive design matters.*

Starting even in preschool, parents and teachers begin to underrate young girls' STEM skills. As early as first and second grade, children can take on their female teachers' stereotypes—both explicit and unconscious—about mathematical aptitude and gender. One study found that while starting in the fall with no bias, by the end of the school year, the more likely the girls were to agree that "boys are good at math, and girls are good at reading," the more likely their math scores also went down.[4] This contributes to a compounding series of small and large obstacles that keep women out of STEM careers, which typically pay two-thirds more than other fields. Today, women make up fewer than 16% of those employed in engineering and architecture roles, and fewer than 3 out of every 10 people working in computer and math jobs are female.[5] *Inclusive education matters.*

In 2014, through something known as gender-balanced budgeting, a group of cities in Sweden[6] reviewed the order in which they plow arterials, side streets, and sidewalks during snowstorms. The analysis found that the practice of starting with streets in employment districts had an adverse effect on women, who were more likely to travel by foot on sidewalks, which were cleared last. What might seem like a hassle—stepping through snow—was actually having a countrywide effect. Nearly triple the number of people were injured falling on ice-covered sidewalks than while driving in snow. So they reversed the order when clearing snow. *Inclusive policymaking matters.*

High-growth startups, the kind that brings disruptive technology to life, account for only approximately 15%[7] of all companies across the United States, but in 2018, the most recent year for which data is available, companies less than a year old created 2.4 million jobs.[8] Put in perspective, this is equivalent to the net change of all jobs, aside from farming, that same year.[9] Unfortunately, companies exclusively made up of women founders receive only 2.3% of all venture capital funding.[10] Assuming that

---

[1] https://www.ncbi.nlm.nih.gov/pmc/articles/PMC3217432/
[2] https://en.wikipedia.org/wiki/Crash_test_dummy
[3] https://deepblue.lib.umich.edu/bitstream/handle/2027.42/1349/94356_rev2007.pdf
[4] https://www.aauw.org/app/uploads/2020/03/Solving-the-Equation-report-nsa.pdf
[5] https://www.aauw.org/resources/research/the-stem-gap/
[6] https://usa.streetsblog.org/2018/01/24/why-sweden-clears-walkways-before-roads/
[7] https://medium.com/@ericcorl/how-startups-drive-the-economy-69b73cfbae1
[8] https://data.census.gov/cedsci/table?q=BDSTIMESERIES.BDSFAGE&tid=BDSTIMESERIES.BDSFAGE
[9] https://www.bls.gov/opub/mlr/2019/article/employment-growth-accelerates-in-2018.htm
[10] https://www.embroker.com/blog/female-founders/

good ideas, talent, and ability to execute are equally distributed across genders, we are collectively missing out on a host of mobility innovations. *Inclusive funding matters.*

I share these examples not as doom and gloom but rather as context for the road we are on. Mobility after all is about movement. I would like to think it is about progress, too. Where we are going matters. On this journey, it is equally important who benefits, who leads, and how we get there. As we have seen in the preceding pages, there is much to be optimistic about. A new category of innovators is changing how we move, and we do not have to wait for futuristic scenarios from sci-fi films to unfold: Mobility's next chapter is already here.

The most obvious characteristic of mobility innovations is that they cause or empower movement or transport. While we often focus on the movement of people, it is important to remember that mobility also encompasses the movement of goods, data, and energy. Sometimes movement, or even lack of movement, in one category replaces another. For example, that next-day package delivery to your home means you did not have to make a trip to the big-box store in the suburbs. Harnessed in new and creative ways, innovative business models, new services, and disruptive technologies open new toeholds of opportunity for startups and create new threats for long-standing industry leaders. Increasingly, too, mobility innovation seeks to drive value through safety, reliability, equity and access, sustainability, and choice.

Finally, we should also remember that the places—those physical domains—where the future of mobility is playing out are not just roads and freeways. Air, waterways, and even space are now home to the future of movement. *Mobility* is truly more than a catchall term for automotive and public transit.

I should pause here and take a step back to provide some context to this point of view on where the future of mobility is headed. In those same presentations that feature the images I mentioned at the beginning of the chapter, I often also start by warning the audience of all of the things I am not. The unexpected break from the list of career bona fides that an enthusiastic moderator has just rattled off catches people's attention and sets the stage for a different type of discussion. Here is why.

I am not an engineer. For long stretches of my life, I have intentionally not owned a car—even in Detroit, the Motor City. I most certainly am not a "car guy." It is also immediately obvious from my gender expression that I am not a guy either. In rooms mostly full of men, calling this out usually gets chuckles from even the most straight-laced, who appreciate the obviousness of the statement.

I see the world of mobility through the eyes of people—its innovators, riders, drivers, students, customers, and passengers. This comes from studying anthropology with its focus on human behavior and society. It comes from a knack for and well-developed habit of perspective taking. It also comes from a love of bike riding, which was my childhood form of freedom. I certainly care how a new technology works or what is innovative about a new approach, but only as a means to an end. How will it be used? Does it solve a pain point and create value? Is someone actually willing to pay for it?

The chance to play a role in the next chapter of Detroit's mobility story is what brought me to this city. First with Zipcar and later with the team at Ford Smart Mobility, I have had the privilege and challenge of working on the frontlines of businesses figuring out how to get new mobility technology and services to market.

My teams have convinced thousands of people to trade the convenience of a car in their garage for the flexibility and savings of car sharing. We have been on the ground for the pivotal moments in mobility's history that nobody will ever write about, including first meetings with mayors to share the idea of launching AVs in their cities. We have built trust with public agencies and forged partnerships that extend the reach of transit systems.

Through initiatives like the Techstars Mobility program, I have mentored dozens of startup founders as they spent summers in the city working to forge connections with local industry contacts and accelerate their chances of success. For startups, the value of building connections to places like Detroit is clear. Unlike any other region in the world, with 24 automotive headquarters and R&D centers in the state,[11] this is definitely the place to build those relationships.

Let me tell you about one other image that further highlights this point. This one is a framed mid-1950s ad for The Detroit Bank[12] hanging in my home office. It was a vintage find that seemed genuinely too good to be true, and I treasure it not only because of my love for mobility and the city but also because of the bold vision captured in this snapshot in time.

In this ad, ships sit moored alongside a Detroit downtown street grid, which, aside from a few notable exceptions, would look unfamiliar to today's residents. Many of the skyline features we know today, including the world-renowned Renaissance Center and the Detroit People Mover, had yet to be built. One of the city's thriving African American neighborhoods was still intact, not yet paved under for a freeway. Look closely enough, and you can actually make out a streetcar heading westward out of downtown. The photo is interesting history, but the headline is great "DETROIT—Birthplace of the Nation's Mobility." Even then, 70 years ago, we realized that transportation is more than movement.

Today, as a mobility entrepreneur and investor based in Detroit, I have co-founded and launched organizations focused on the future of people and place within global mobility. At Assembly Ventures we believe firmly that while innovation is global, mobility is local. Digital products can become accessible anywhere, and an international approach can maximize market share. Yet movement requires a physical presence, and for startups, this extends to proximity to legacy industrial centers. Bridging this gap is central to our strategy, and I would argue will also be a hallmark of Detroit's future.

Also key to this future is talent. Research we conducted in 2018 at the Michigan Mobility Institute in partnership with the Boston Consulting Group highlighted the urgency of the situation. Through dozens of interviews with leaders across the mobility landscape, we recognized that sustaining and growing the mobility industry's current momentum also required honing the approach to education. By bringing the industry more closely alongside education partners, we believed we could get people ready with the skills and experience most in demand today. Across the United States, we will need more than 100,000 new mobility technicians and engineers in the next decade.[13]

---

[11] https://michauto.org/data-highlight-michauto-automobility-asset-map/
[12] https://www.linkedin.com/pulse/detroit-birthplace-nations-mobility-jessica-robinson/
[13] https://www.bcg.com/press/11january2019-mobility-and-automotive-industry-create-jobs-exacerbating-talent-shortage

In one of the most critical areas—software engineering—the industry will need six times as many people with computer-focused degrees versus those actually graduating. Ramping up the talent pipeline, including engaging a diverse group of workers, is essential for the industry's near-term viability.

One of the hallmarks of a mobility engineering mindset is comfort with and acceptance of change. Working within mobility is a paradox, planning for what you can't quite imagine. Who would have thought that fleets of low-cost scooters would burst open a new category of shared micromobility, simultaneously democratizing access for some, reinforcing barriers for others, and creating trash heaps of discarded units? Or that a global pandemic would pause daily commuting for entire categories of workers and drive mass experimentation with videoconferencing for everything from meetings to birthday parties? The ability to create, spot, amplify, and invest in emerging trends with the right timing means everything.

Today's modern transportation solutions span modes and seek to offer customers the right choice at the right time in a way that is increasingly responsive to their changing needs. Yet we must also recognize that the systems themselves are in constant transformation as they are exposed to social, economic, and regulatory influences. Urged forward by increasing connectivity and digitalization, mobility options will continue to become multimodal, autonomous for certain applications, and highly flexible. Reduced or eliminated silos between systems and industries will further improve efficiencies.

While the physical world plays a central role within mobility, infrastructure innovation has lagged. This can be attributed to many causes, including long investment timelines, federal-state funding structures, and complex environmental reviews. We must think through which parts to preserve and which to let go. In some cases, infrastructure innovation requires the opposite, rallying behind large-scale, transformative projects to launch new systems.

This is not meant to imply that future-proofing mobility infrastructure will be easy. Cities and public transit agencies must balance operating legacy infrastructure for current systems that are not optimized let alone optimizable for new technology or modes. Before the COVID-19 pandemic, many were already at or beyond their breaking point. However, in remaining stuck within the same ways of building and maintaining roads, rail, and bridges, we risk creating a generation-long drag on mobility innovation.

This physical infrastructure was built to serve the singular function of facilitating movement. Getting change right will be difficult enough. So too is the case with energy infrastructure, which by necessity must support diverse needs.

In the United States, the transportation sector consumes more energy than any other. It first surpassed the industrial sector in the 1980s and has continued an upward trend since then. Both transportation and industry far outpace the energy required by residential and commercial properties. Today, this demand is fed primarily by oil. In fact, 69% of our total transportation energy comes from petroleum or natural gas.[14] As EVs become increasingly available, demand will shift to electricity and place new burdens on an already strained grid.

---

[14] https://www.eia.gov/totalenergy/data/monthly/pdf/flow/css_2019_energy.pdf

Public utilities, facing this surge of electrification and the prospect of adding a new moving layer to their energy systems, will need to get ahead of a new set of challenges. They will need to quickly figure out how to bring new power densities to charging depots for electrified fleets. Importantly, these depots may not share the same neighborhood footprint. Public transportation yards located in city centers may be miles away from the suburban distribution centers where private delivery trucks are staged. Utilities will need to find new storage solutions to smooth out the availability cycles of renewables like wind and solar. They will need to come up with smart tools and pricing mechanisms to nudge consumer behavior and shift personal charging to off-peak hours. Perhaps most challenging, public utilities will need to convince their regulators that proactively expanding charging networks, offering new services, and investing in new technology is a worthwhile effort.

In fact, for our physical transportation infrastructure and our energy infrastructure, advancing these legacy networks requires reimagining use and making selective enhancements to overlay digital technology onto existing "dumb" asphalt, steel, and transmission lines. Hidden behind the scenes, new digital systems play a key role. They coordinate activities by unlocking and enabling new flows of data and ensuring these transmissions are secure from malicious hacking. Through their structure, these systems collect and store new forms of data and own the rules about whether it can be accessed through open protocols or will remain locked in a proprietary network. Complex algorithms, fed by this data, manage and direct decisions as varied as the path an AV chooses to the pricing of airline tickets based on trending website activity.

While it remains true that foundational infrastructure and the physical-digital interface are primarily about capacity, at the end of the day mobility is about demand. These enablers connect us with entire categories of new mobility offerings in a dizzying array that ranges from shared rides, cargo matching, and mobility-as-a-service smartphone apps. Here flexibility, sustainability, access, reliability, and safety drive customer choice.

As activities approach a post-pandemic norm, mobility consumers will look for ways to return spontaneity and flexibility to trips, to facilitate contactless transactions, and to help balance changed work-life demands.[15] Automotive manufacturers and suppliers, reeling from more than a year of compounding supply chain disruptions due to the pandemic and semiconductor shortages, will also prioritize flexibility. They will also need ways to build in new resiliency and efficiency while protecting against the exposure of ultra-just-in-time supply. Public transit agencies will explore deeper partnerships with the private sector in an effort to preserve overall network service while focusing operations on core routes.

As we continue this massive shift from automotive transportation to broader mobility, let us not forget that getting an entire set of industries to collectively move in a common direction is no small feat. This is not about getting one elephant to dance. It is getting the entire menagerie at the zoo to come out of their cages and line dance together. Real investments in R&D and the commitment to commercializing

---

[15] https://www.businesswire.com/news/home/20210118005006/en/Top-10-Global-Consumer-Trends-in-2021

new forms of technology are essential. The will to invest and risk potential failure is required. Incentives and policy help. I would also argue that inspiration is key.

In anthropology, we have the concept of *myth*. The idea certainly includes the common definition of sacred stories, memorable heroes, and supernatural explanations. However, there is a broader concept that applies within the business and to this particular moment in mobility's evolution. It is the idea that as people, we use a collection of stories and narratives to explain where things come from, why we are here, and where we are going.

In getting to this milepost on the mobility industry's journey, we have largely relied on a framework described by some version of the acronym CASE. In doing so, we have collectively recognized and rallied around the roles that connectivity, autonomy, shared mobility, and electrification will play in shaping consumer expectations and driving innovation. The framework has been essential for companies puzzling through how they get from where they are today to where they need to be in the future. CASE has been mobility's myth—to this point in time.

While powerful as an organizing principle, CASE cannot be relied upon to predict the future. Indeed, the framework fractures depending on which side of an equation one sits on. If a company focuses primarily on designing and producing vehicles, it might talk more about CAVs or vehicle-to-infrastructure (V2I). Doing so lets it keep the car central to the story. The flipside is also true. If a company works on the connectivity side of the equation, it might refer to the internet of things (IoT). We even have pedestrian-to-everything (P2X) and pedestrian-to-vehicle (P2V) for advocates and others focused on vulnerable road users.

Lost in the acronym soup is a deeper truth: innovation in autonomy, electrification, and connectivity are concerned with bridging a divide between our physical and digital worlds to enable human aspiration. Truly unpacking what this means for mobility has the potential to catalyze new alignment about why we are doing any of this work in the first place. If our contemporary CASE myth focuses on the *what* and *how*, our new narrative should shift to *who* and *why*. This is essential not just for the overall advancement of the industry but also because myth is deeply linked to who participates in the storytelling. When we shift beyond CASE, we put people back at the center of mobility choices. Importantly, given the context of this book, we also set the table for real gender inclusivity.

So where do we go from here? And just as important, with whom will we choose to travel on this journey?

While bringing women into all levels and roles of the mobility sector is a start, it is an incomplete answer. As a (female) executive at an AV startup recently reminded me, there is a difference between outputs and outcomes. What we choose to measure matters. Within that choice, the following are a few ideas for how to accelerate from here.

Make female role models more visible within mobility.
You have now met dozens of innovators who fall into this category. Sharing these stories, particularly with young students, combats gender stereotypes and highlights the ways in which different backgrounds are valuable within the field. Finding ways to connect efforts like this book or the *Remarkable Women in Transport* compilation to in-school or worksite visits can help girls imagine their futures in mobility.

Support those mentors, too.

Today's women leaders in mobility and transportation are carrying a heavy burden. Well-known challenges of work-life balance and the decision to raise a family influence not just how far women rise in their careers, but also sentiments about the work they have chosen in this field. A 2019 survey[16] of women in the automotive industry found that 47% of respondents would think about a different industry if they were starting their careers again. This type of disillusionment, extended to an industry like transportation in which women make up less than 15% of the total workforce, matters. Seemingly straightforward policies and practices like actually making sure people take earned vacation days and allowing for ample time off following childbirth[17] can improve recruiting, job satisfaction among current employees, and long-term retention.

Celebrate women in the boardroom and seek all forms of diversity.

Among the 3,000 largest U.S.-traded stocks, women now hold 20% of board seats,[18] and the rate of companies growing their number of board seats to add women directorships is also increasing. Notably, just as this book was being finished, GM became the only U.S. automaker to have a majority-women board.[19] Widening searches to include age, race, and professional background[20] with the goal of expanding a board's overall diversity can improve culture, the way decisions are made, the ability to monitor the company's managers, and its overall knowledge base.[21]

Find, support, and fund women-led mobility startups.

Startups increasingly play a key role in driving mobility innovation forward. In 2020, mobility ranked as the single largest sector among the top 100 startups to receive funding in Germany. Collectively the segment saw an increase of 36% in 2019. While only 14 of the 100 total were co-led by women, headwinds are slowly turning into tailwinds for female mobility co-founders and early hires.[22] Programs like Empower Women in Shared Mobility and MobilityXX are intentional ways to source and support women-led startups. Continuing and expanding these types of efforts create a double win by finding great companies and building cohorts of entrepreneurs who can act as mentors to one another as well as the next generation.

Invest in better public transit.

Public transit is an essential backbone of a passenger mobility ecosystem. While underfunded and under-resourced, these systems are a lifeline for many, and the only way to get around. During the first peak of the coronavirus and the associated state-wide stay-at-home orders, Transit app found that among their user base in North

---

[16] https://www2.deloitte.com/us/en/insights/industry/automotive/women-in-automotive-sector-gender-diversity.html
[17] https://transweb.sjsu.edu/sites/default/files/1893-Godfrey-Attract-Retain-Women-Transportation.pdf
[18] https://2020wob.com/wp-content/uploads/2019/10/2020WOB_Gender_Diversity_Index_Report_Oct2019.pdf
[19] https://www.reuters.com/article/us-general-motors-directors/women-hold-majority-of-seats-on-gms-expanded-board-of-directors-idUSKBN2BH1X3
[20] https://hbr.org/2019/03/when-and-why-diversity-improves-your-boards-performance
[21] https://corpgov.law.harvard.edu/2020/07/14/maximizing-the-benefits-of-board-diversity-lessons-learned-from-activist-investing
[22] https://assets.ey.com/content/dam/ey-sites/ey-com/de_de/news/2021/03/ey-venture-capital-study-tech-startups-2021.pdf

America,[23] ridership shifted from 50/50 to 56% women. Critically, they also saw half of white passengers drop out of the system, leaving primarily Black and Latina women riding. Committing to sustainable funding models and creative private partnerships can ensure that we preserve transit-enabled mobility, which is essential for women.

Embrace human-centered design.

Building for experience, not just efficiency or capacity, can make mobility more accessible. Motivated by safety concerns, women have been found to bear a "pink tax,"[24] paying $26-$50 more than men on a monthly basis for transportation by choosing private options over public transit. Designing subway platforms around safety, for instance, could reduce this burden.

In offering these final thoughts for the road ahead, I certainly do not claim to have a crystal ball. Still, I do believe that someday soon, when I look out into the room at a conference and share those same opening photos in my slides, the faces sitting in the crowd looking back at me will be different. I will still see those car guys. But sitting beside them will be women like Dr. Tierra Bills, looking for ways to improve access to microtransit; women like Pam Fletcher, leading the transformation of a global automaker from the inside the trenches; women like Jamie Junior, raising her voice to make sure people with disabilities are included in mobility planning; and women like Lisa Nuszkowski, conjuring bikesharing to life in the Motor City.

If we are wise, we will have used technology alongside inclusive design, education, policy, and funding to deliver on mobility's deepest aspiration and myth: freedom.

---

[23] https://medium.com/transit-app/whos-left-riding-public-transit-hint-it-s-not-white-people-d43695b3974a

[24] https://newcities.org/the-big-picture-the-pink-tax-on-transportation-womens-challenges-in-mobility/

About the Authors
The Mobility Industry's Storytellers

Photography by Elizaebth Moroz with expdet.com

There is no greater agony than bearing an untold story inside you.

—Maya Angelou

After our professional circles narrowly skimmed past each other for years, our paths finally collided in January 2019 on a cold, snowy day in downtown Detroit. It was the third day of the North American International Auto Show (NAIAS), an event the two of us had heavy involvement in for quite some time. Katelyn was working as the director of the Michigan's automotive and mobility industry association,

MICHauto, and Kristin was working as Digital Marketing Manager for TCF Center (formerly Cobo Center), the convention center that had been home to NAIAS for more than 50 years.

There began both a friendship and a professional pairing between Katelyn and Kristin, where we quickly bonded via Twitter over our love for the show's excitement and prestige and ended up meeting up for a behind-the-scenes tour of the convention center led by Kristin for Katelyn' social media takeover, aimed at marketing NAIAS. We spent the morning coming up with story lines that would attract people to the show and create a global lure for the vehicles it debuted and the thought leaders who took the stage. We told the story of Detroit's global leadership in automotive and its commitment to reinventing the way the world moves through a new era of mobility. After that, you might say, the rest was history.

Once a storyteller, always a storyteller.

Later in 2019, our paths crossed again. This time Katelyn had gone back to her public relations roots and was working at an agency that specializes in transportation and mobility and Kristin was working at WSP, one of the world's leading engineering consulting firms, as Michigan's Communications and Public Involvement Coordinator. We had been asked to join the Women in Mobility-Detroit's operations team. This group, started in 2018, had become the premier industry group for the international mobility community based out of the world's automotive hub and historic "Motor City," Detroit. Though the organization was called "Women In Mobility," the mission was to improve gender diversity for not just women but womxn. This important distinction led Katelyn, Kristin, and the operations team to reconsider the audience, goals, and outreach efforts to make sure that the door was open to anyone, regardless of how they identify. It became very clear to us that the gaps in gender representation did not start and begin with women or with the feminine identity. It just seemed to start and continue within the limits of a masculine perspective. Once again, our proclivity for storytelling was put to work.

We went to work with the operations team to formalize the group's operations and to increase awareness by getting the message out more broadly in the local community. Over the course of a year, Women in Mobility-Detroit launched a website and a virtual series of events (due to COVID-19) and increased its membership by more than 300%. By curating events with thought leaders who resonated with the community, we—along with operations committee members Caroline Choudhury, Stacey Matlen, and Hind Ourahou—created a platform that elevated female voices and allowed members to share best practices, receive mentorship and support, and learn and grow within the industry, as well as encourage more women to join this underrepresented sector. This organization facilitated an international partnership between Detroit and the Windsor-Essex region of Ontario, Canada—with help from Nicole Anderson of the Windsor-Essex Small Business Centre and Yvonne Pilon of WEtech Alliance—located directly across the Detroit River in Canada. Together, we developed a narrative around the binational economic development and talent opportunities in automotive and mobility.

Despite all the work we were doing in our small region, we were seeing the gaps in female leadership and representation in the mobility industry at every turn. The representation of female speakers at events was low, the companies and stories featured in the media were more than likely represented by male associates, and operations were male dominated. We had the utmost pleasure of surrounding ourselves with women who were pioneering technical advancements, designing innovative transportation systems, and changing the way people, goods, and data move around the world—many of whom are featured in this book—and we knew the industry could do better.

We felt at first like it was not an optimal time to write a book, but honestly, it was the perfect space to allow us to absorb the events of everything pre-pandemic along with the historic events of 2020 and to really imagine what the world would look like after the pandemic was over. Though there has been a significant amount of grief, mourning, and adjustment during the writing process of the book, the moment felt right to tell this story. Much of the industry has kept advancing while other areas have pivoted, but regardless of each individual response to the "new normal," one thing was certain: *we could not go back to the way things were.* This moment provided a unique intersection for us to work together with SAE International to help tell these stories, to showcase the incredible work of women across the United States, and to inspire the next generation. Though it has felt like we are wearing many hats professionally, unique professional portfolios and experiences positioned us to turn the volume up on this conversation by amplifying the stories included in *Women Driven Mobility*.

Katelyn Shelby Davis

Katelyn Shelby Davis serves as the Mobility Lead at Antenna, an integrated communications and marketing firm, where she provides wraparound public relations services to global clients within the automated, electric, and connected vehicles; mass transportation systems; and micromobility ecosystems. For the past decade, she has been ingrained in the automotive and transportation industry, including her role as Director of MICHauto, Michigan's automotive and mobility industry association, where she developed programs that helped promote, retain, and grow the state's signature industry.

Growing up in the Motor City, her love for automobiles started at a young age and has influenced her career as a passionate storyteller for some of the world's leading mobility companies. She has taken that passion and assisted automakers, suppliers, startups, community organizations, and government entities in developing better and more efficient ways to move people, goods, and services.

Davis is an award-winning communications professional, recognized with six national titles for strategic communications campaigns. She was named one of Detroit's 30 in their Thirties by *DBusiness* in 2021. She is an active board member of the Automotive Public Relations Council and is a graduate of Leadership Detroit. Formerly, she served as Editor in Chief of *Driven*, a news outlet showcasing Detroit's global leadership in mobility, and as part of the Women in Mobility-Detroit board. She also served as a Girl Scout leader, committed to mentoring the next generation of female leadership. She is a graduate of Grand Valley State University and Wayne State University, where she studied advertising and public relations, as well as new media communications.

Davis lives for the intersection of technology and transportation and the creation of equitable, accessible, and sustainable transportation that serves all.

Kristin Shaw

Kristin Shaw works in communications and public involvement at WSP, one of the world's leading engineering professional services consulting firms, where she is responsible for outreach and public involvement efforts for some of the largest and most innovative transportation, resilience, and infrastructure projects across the United States. Formerly, Shaw served as Digital Marketing Manager for TCF Center, Detroit's Convention Center, where she also served as the project administrator for the 2019 LEED Gold Certification and oversaw many of the facility's corporate social responsibility programs and sustainability efforts.

As a versatile marketing and communications leader and photographer, Shaw draws from a passion for visual storytelling, sustainability, and urban planning to focus on transportation systems that reduce emissions and provide more equitable and accessible options for all users. With a focus on human behavior and experience, Shaw believes that the best places are built from the collective imagination. She has significant experience in both policy development and implementation regarding materials management, green buildings, sustainable business practices, environmental education, and marketing, as well as working with environmental justice communities.

Shaw serves on the Michigan Leadership Advisory Board for the USGBC Detroit region and as a co-chair of the Detroit City Council Recycling and Waste Reduction Committee. She is a member of the American Planning Association, Women in Transportation, and Women in Climate Tech and has been involved with the Midtown Cultural Connections planning team in Detroit. Shaw additionally works with the State of Michigan Council on Climate Solutions: Buildings and Housing and Transportation and Mobility Working Groups Shaw has received numerous awards and accreditations throughout her career. She is a LEED-accredited professional and was recognized with the 2019 USGBC Detroit Outstanding Leadership in Sustainability award, Michigan Association of Planning Outstanding Graduate Project award, Detroit Young Professionals Vanguard award, Public Relations Student Society of America Brick award, and Wayne State University Alumni Award, and she was honored in 2019 to be among *Crain's Detroit Business* 20 in Their Twenties. Shaw is a featured artist in the 2021 SaveArtSpace exhibit "At Whose Expense," where her combined passion of design and storytelling come together to press the issue of a climate emergency. In addition, Shaw volunteers with the Detroit 2030 District, Green Living Science, EcoWorks Detroit, Bees in the D, and other local organizations. Shaw holds a bachelor's degree in Public Relations and a master's degree in Urban and Regional Planning from Wayne State University.

Shaw believes that with the right amount of listening, learning, and action, equitable and sustainable cities are just around the corner.

Special Thanks

Writing a book is a rewarding, yet exhausting, process. We have been lucky to have each other to lean on during the experience; however, this book would not have been possible without the support of so many people—namely, our publishing team, supportive bosses and teams at the office, our friends, family members, and the many people we met along the way. This has truly been a collective accomplishment.

Thank you to Sherry Nigam, Linda DeMasi, and Bruce Sherwin for your support and sponsorship of this important topic and including it on your esteemed list of publications at SAE International.

Thank you to *That Woman in Michigan* Governor Gretchen Whitmer for your leadership over our great state; your dedication to a stronger, greener, and more connected transportation network and economy; and inspiring us to stand up and use our voice. Furthermore, we thank the Governor's team, Michelle Grinnell at the Michigan Economic Development Corporation, and Trevor Pawl from the Office of Future Mobility and Electrification for working with us on this book.

Thank you to our artist, Quinn Faylor, for your commitment to equity, storytelling, and creativity and bringing the vision of this book to life with your art, and to Danielle Shields for the introduction.

Thank you to Jessica Robinson for all your input into this book, your guidance along the way, and everything you do within this community as a role model for other women. There's no one better to be representing on us all on those stages full of men.

Special Thanks

Thank you to everyone who made introductions, passed along leads, shared articles, and pointed us in the right direction, especially Kelley Coyner who spent endless hours making introductions and summarizing the work of dozens of women.

Thank you to the many men who fight for women's seat at the table, make room, share the stage, and are first to move over for somebody else. We see you, we appreciate you, and the wins in this book were made in your company. We have made it a piece of our journey as authors on a gendered subject to recognize that greatness cannot come from one or even two genders, but all.

To the many organizations who helped tell this story and wonderful individuals that make these organizations impactful: Caroline Choudhury, Hind Ourahou, Stacey Matlen, Janine Ward, Tara Lanigan, and Anne Partington with Women in Mobility; Lana Crouse at the United States Green Building Council; Katie Kennedy and the team at EXP|DET; Lisa Nuszkowski and Adriel Thornton at MoGo; and Jay Pitter. And thank you to the many people who supported the projects listed in this book: Dr. Tierra Bills, Pam Fletcher, Jamie Junior, Gayle Schueller, Cristina Thomas, Paul Acito, Ellen White, Sinan Yordem, Justin Johnson, Jonah Shaver, Bob Anderson, Ben Watson, Andy Ouderkirk, Kevin Plesha, Gerrad Bailey, Ken Smith, John Wheatley, Joline Bogdan, Dave Meslow, Cris Asuncion, Meredith Moore Crosby, Randy Iwasaki, Pam Henderson, Tammy Meehan Russell, Jay Hietpas, Kristin White, Kian Sabeti, Janelle Borgen, Daryl Taavola, Mike Kronzer, Cory Johnson, Laurie McGinnis, Phil Magney, Patrick Weldon, Susan Mulvihill, Margaret Anderson Kelliher, and Myrna Peterson.

Kristin extends special thank yous: To Dr. Shelly Najor for being a rock on so many occasions, a mentor, and one of the strongest and most fun woman she has the pleasure to know. This book wouldn't exist if I hadn't learned from you. To my team at WSP: Shane Peck, Scott Shogan, Dr. Rawlings Miller, Denise Turner-Roth, Yosef Yip, Melissa Uland, Alyssa Curran, and Emily Wasley for your work in the industry and your support and involvement as this project developed. To Lisa Hennessy, Gina Baker, Greg DeSandy, Jennifer Berkemeier, Claude Molinari, Thom Connors, and Cedric Turnbore for believing in me early in my career. To my dear friends for support over the phone and on Zoom and for your patience while we worked on chapters during the time we were able to spend together last year—Amber Devlin, Andrea Shaw, Evelyn Shea, Jane Pawelski, Kate Sprader, Kayla McRobb, , and Natalie Jakub— and to Jessica Sader, thank you for your friendship and being the best copywriter I have ever met and your willingness to always help with all things about writing. To Brice Moss, Richie Garcia, Ben Dueweke for championing women and always having my back. To my Aunt Kae for making sure I was never on my own. To my siblings, Jenna, Audrey, Mitchell, and Issaac, for teaching me how to fight for the silenced, the vulnerable, and for sharing in our power to overcome adversity. To my dad Joseph Shaw for reminding me that there are still many men who can love women with their whole heart, and still have so much to learn—for teaching me to be independent and to be an ongoing reminder that forgiveness is more powerful than pride. To my mom, Debbie Leidner, thank you for passing down your laugh to me, your creativity, and your fight, shared to me with so much love. I am lucky to be your daughter.

Katelyn extends special thank yous: To my parents, Mark Davis and Linda Tarolli, for which I owe everything. It was all those long weekends at classic car shows and car cruises with the family that jump-started my love for automotive, and it was your love, unwavering support, and insistence that I could do anything—that kept me going for all these years. To my current and former colleagues: Misty Matthews for your guidance and mentorship so early on in my career, Glenn Stevens for your support and friendship always, Samantha Roberts for helping me to grow as a leader, and Lisa Lark for your insight and advice as a fellow author and friend and for being a sounding board throughout the last two years. To the many members of the Leadership Detroit Class XLI for your encouragement along the way. To Carlotta Gmachl, Vittoria Valenti-Amodeo, Kathryn Snorrason, Jenny Orletski-Dehne, Kate Partington, Kathryn Smith, Tabitha Mensch, Marnita Harris, Katie Krizanich, and Amanda Roraff for being a part of some of my favorite experiences in this industry and for making sure none of us ever felt alone.

Last, but absolutely not least, we extend our greatest gratitude to our partners, Marc Langlois and Mike Bargerstock, and our pups, Juneau and Margo, for making sure we had dinner and found time to sleep, celebrated our wins, and set the book down when we needed a walk outside.

About the Cover Artist

Quinn Faylor is a queer, nonbinary multidisciplinary artist living and working in Detroit, Michigan. They were born and raised in Petoskey, Michigan, and received their B.A. in Arts and Ideas in the Humanities from the University of Michigan in 2016. Since finishing their undergraduate degree, they have completed a thru-hike of the Appalachian Trail and Colorado Trail..

Their work centers on embodiment and joy. The content of their work spans from the highly personal, intimate oil portraits of the human body to the abstract, large-scale acrylic murals that live in public and private residences.

No matter the content, their compositions are intended to be a celebration. The colors feel exuberant. Each composition is grounded in a series of bold and contrasting shapes. They enjoy layering colors on top of each other and maintaining distinct boundaries between them. The shapes carry motion. Their work is directional. It brings the viewer into and through a space.

At its best, their work is joyful. Their practice is an extension of themself.

Website: deliciousgoldmakes.com.

Index

A

"Accessible, Automated, Connected, Electric, and Shared" (A2CES) mobility future, 149–150
Advanced air mobility (AAM), 185
AirHub™, 186–188
Airspace Link, 186–188
   advanced air mobility (AAM), 185
   approach and challenges, 186–187
   FAA approval, 187
   goals and objectives, 186
   large-scale commercial drone operation, 185
   UAVs, 185, 186
American Association of State Highway and Transportation Officials (AASHTO), 9, 11
American Automakers Policy Council (AAPC), 120
American Community Survey (ACS), 9
Americans with Disabilities Act (ADA), 24
Arens, Barbara, 64–65
Attarian, Janet L., 114–115, 179–180
Automatic wheelchair securement system, 25
Automotive industry
   decision-making authority, 12
   female gender diversity, lack of, 12
   women in, 2, 12
Autonomous vehicles (AVs), 1
   education and outreach, 8
   May Mobility, 24–25
Awareness and community advocacy
   Census Transportation Planning Products Program (CTPP)
   approach and challenges, 10
   baseline dataset, 10
   commuter flows from home to work, 9
   goals and objectives, 10
   nation's workforce mobility, 10
   nontransportation industries, users in, 10
   reports and research papers, 11
   residence-based data, 9
   transportation planning community, 9
   workplace-based data, 9
EV Workforce and gender gaps, ownership
   approach and challenges, 13
   decision-making authority, 12
   goals and objectives, 12
   income-based disparity, 12
Michigan long-range transportation plan public engagement partnership
   approach and challenges, 18–19
   FHWA policy, 15
   goals and objectives, 17
   Michigan Mobility 2045 (MM2045), 15, 17
   mission statement, for plan development, 16
   public engagement strategy and execution, 16
   state DOTs, 16
   State public involvement and public hearing procedures, 15–16
   platform modifications, 10
   public transit systems, 8

B

Baxter, Margaret, 165
Begalli, Adelaide, 27, 28
Black, indigenous, and other people of color (BIPOC), 134

Boston Consulting Group
(BCG) study, 40
Busino, Laura, 88
Bus Rapid Transit (BRT), 46
C
California Air Resources
Board (CARB), 70, 71
Census Transportation
Planning Products
Program (CTPP)
approach, 10
baseline dataset, 10
challenges, 10
commuter flows from
home to work, 9
goals and objectives, 10
nation's workforce
mobility, 10
nontransportation
industries, users in, 10
reports and research
papers, 11
residence-based data, 9
transportation planning
community, 9
workplace-based data, 9
Center for Automotive
Diversity, Inclusion,
and Advancement
(CADIA), 162
Clean energy technology, 13
Committee on Data
Management and
Analytics
(CDMA), 11
Community Organized
Relief Effort
(CORE), 203
Connectivity, Autonomy,
Shared mobility
and Electrification
(CASE) technology,
183, 184
COVID-19 pandemic
Detroit, COVID-19
transportation
response
approach and
challenges, 204–205
DDOT, 204

goals and objectives, 204
Office of Mobility
Innovation (OMI),
204, 205
mobility venture
incubator
approach and
challenges, 202–203
goals and objectives,
202
public health crisis, 198
safe transportation, 198
"stay at home" order, 198
tribal communities
transportation
assistance, in remote
communities
approach and
challenges, 200
goals and objectives,
199–200
GoFundMe
campaign, 200
Navajo Nation, 199
White Mountain
Apache
community, 199
Coyner, Kelley, 152
Culler, Mary, 109
D
Datta, Bonnie, 72–73
Daynamica, smartphone-
human hybrid
intelligence system
approach and
challenges, 29
biking-based
transportation
happiness map, 32
bus-based transportation
happiness map,
32, 33
customer service
outreach, 28
driving-based
transportation
happiness map,
30, 31
emotional landscape, 28
goals and objectives, 29

human-centered
transportation
planning
practices, 28
Minneapolis-St. Paul
Transportation
Happiness Map,
32, 33
Departments of
Transportation
(DOTs), 9
Design and engineering, of
automobiles
accessible AV
Transportation, for
those with
disabilities/
impairments and
seniors
approach and
challenges, 24–25
driverless
transportation, 23
goals and objectives, 24
level 5 self-driving
vehicles, deployment
of, 23
transportation
independence, 23
crash-test safety
protocol, 22
gender diversity in urban
design, lack of, 22
military vehicles, 21
revolutions per minute
(RPM) to
experiences per mile
(EPM)
approach and
challenges, 36
EPM Advisory
Council, 36–38
goals and objectives, 36
HARMAN and SBD
Automotive, 35, 36
smartphone-human
hybrid intelligence
system, measure and
shape happiness,
2020

approach and
challenges, 29
biking-based
transportation
happiness map, 32
bus-based
transportation
happiness map, 32, 33
customer service
outreach, 28
driving-based
transportation
happiness map, 30, 31
emotional landscape, 28
goals and objectives, 29
human-centered
transportation
planning practices, 28
Minneapolis-St. Paul
Transportation
Happiness Map, 32, 33
sustainable materials,
fashionable and
handcrafted interiors
approach and
challenges, 27
goals and objectives,
26–27
Revero GT SCI concept
car, 27
Design thinking, 3
Diverse representation, 4

E

Electric Avenue
approach and challenges,
142
goals and objectives,
140–141
sustainable
transportation, 140
Environmental Protection
Agency (EPA), 18

F

Fan, Yingling, 34
Farnsworth, Elaina, 174–175
Federal Highway
Administration
(FHWA)'s
requirements, long-
range planning, 15

Female leadership, 4
FlexLA microtransit system,
96–98, 100
Flynn, Jacques, 27
Ford Motor Company, 40
Ford X, mobility venture
incubator
approach and challenges,
202–203
goals and objectives, 202
testing and vaccines,
mobilization of, 203
Funding
COVID-19 pandemic, 40
economic development
approach, funding
mobility projects
(2018–2020)
Airspace Link, 43
approach and
challenges, 42
Derq, 43
goals and objectives, 42
PlanetM Mobility
Grant program,
41, 42
Pratt Miller, 43
Propelmee, 43
female-led funds,
lack of, 40
gender gap, 40
Regional Transit
Authority (RTA),
Southeast Michigan
Ballot Initiative
approach and
challenges, 47
goals and objectives,
45–47
mission, 45
transit millage, 48
transportation master
plan, 46, 47
safe pedestrian and bike
routes, Tribal Land,
Karuk Tribe
approach and
challenges, 51–52
goals and objectives,
50–51

Native American
community, 49
Technical Advisory
Committee
(TAC), 53
Tennessee corridor fast-
charging network
approach and
challenges, 55–56
goals and objectives,
54–55
Tennessee Department
of Environment and
Conservation
(TDEC), 56
Tennessee Valley
Authority
(TVA), 56
Venture Capital
(VC), 40

G

Gender diversity, 3
General Motors (GM), 2
approach and challenges,
82–83
goals and objectives, 82
Super Bowl, 83
Geographic Information
System Mapping
(GIS), 10
Gillis, Greer Johnson, 69
Global positioning system
(GPS), 189
Grellier, Penny, 93
Guerrero, Dibrie, 88
Gunter, Tiffany, 48–49

H

Hanson, Melinda, 143
Healander, Ana, 188
Hennepin County
Community Works
approach, 116

I

Infrastructure
decarbonized
transportation and
mobility system, 60
Interstate 75 (I-75)

approach and
challenges, 62–63
Diverging Diamond
Interchanges
(DDI), 61
goals and objectives, 62
High-Occupancy
Vehicle (HOV) lane,
61, 62
safety, 62
segment 1 work, 63
segment 2 work, 63–64
segment 3 work, 64
mobility optimization
through vision and
excellence (MOVE)
approach and
challenges, 68
goals and objectives,
66–67
Jacksonville
Transportation
Authority (JTA),
66–67
roadmap, 68
open access electric
vehicle charging
approach and
challenges, 70–71
charger anxiety, 70
goals and objectives, 70
greenhouse gas (GHG)
emission
reductions, 69
radio-frequency
identication
(RFID), 70
Internal recognition, 4
Internet of things
(IoT), 36

J

Jacksonville Transportation
Authority (JTA),
66–67
Journey to Work (JTW), 9

K

Karma Automotive, 26
approach and
challenges, 27
goals and objectives,
26–27
Revero GT SCI concept
car, 27
Karuk Tribe, Happy Camp
community and
approach and challenges,
51–52
characteristics, 50
goals and objectives,
50–51
Native American
community, 49
Technical Advisory
Committee
(TAC), 53
Kukreja, Anusha, 196

L

Lanigan, Tara, 26
LEED for Cities pilot
goals and objectives, 144
LEED v4.1 for cities and
communities,
144–145
local governments,
impact of 2020 on,
145–146
STAR Community
Rating System
(STAR), 143, 144
urban sustainability,
146–147
Light detection and
ranging (LIDAR)
sensors, 189
Light Rail Transit (LRT)
line
approach and challenges,
116–117
goals and
objectives, 116
locally preferred
alternative (LPA),
115
transit-oriented
developments
(TODs), 117
TSAAP process, 117
LTRP strategies, 19
Luis, Andre Franco, 27

M

Malek, Alisyn, 126
Marketing and
communications
autonomous and
connected
vehicles, 76
Bronco
approach and
challenges, 79–80
Ford Motor Company's
trajectory, on future
vehicle models, 77
goals and
objectives, 79
off-roading
segments, 81
GM
approach and
challenges, 82–83
goals and objectives, 82
Super Bowl, 83
multicultural marketing
"Built
Phenomenally"
team, Ford Motor
Company
African American
female audience,
85
approach and
challenges, 85–86
decision-makers, 87
Matlen, Stacey, 206
May Mobility, 24–25
approach and challenges,
24–25
driverless
transportation, 23
goals and objectives, 24
level 5 self-driving
vehicles, deployment
of, 23
transportation
independence, 23
McCurry, Erin, 25–26
MEP Mobility Hub/THE
HUB
approach and challenges,
176–177

goals and objectives,
175–176
Mercy Education Project
(MEP), 175
Meunier, Nicole, 178
Michigan Department of
Civil Rights, 19
Michigan Department of
Transportation
(MDOT), 15–19,
61–63
Michigan Economic
Development
Corporation
(MEDC), 41
Michigan Long-Range
Transportation
Plan, 15
Michigan Mobility 2045
(MM2045), 15
Mobility innovations, 7.
See also Technology
innovation
Mobility on Demand (MoD)
Bedrock's parking
facilities
approach and
challenges, 94–95
goals and objectives, 94
MyCommute
platform, 95
stable, reliable and
stress-free parking
platform, 94
transportation,
alternative modes
of, 95
FASTLinkDTLA
approach and
challenges, 99–100
data collection, 98
goals and objectives,
98–99
traffic congestion, 98
transportation blank
spots, 100
Limited Access
Connections
approach and
challenges, 92

goals and objectives,
91–92
Pierce Transit, 92–93
single-occupancy vehicle
(SOV) trips, 90
USDOT, 90
user-centric approach, 90
Monsma, Monica, 19
Myers, Erika, 13–14

N
National Association of
City Transportation
Officials
(NACTO), 143
National Automobile
Dealers Association
(NADA), 22
National Environmental
Policy Act (NEPA)
process, 15
North American
International Auto
Show (NAIAS), 76
Norton, Hilary, 101

O
Office of Future Mobility
and Electrification
(OFME), 43
Original equipment
manufacturers
(OEMs), 36
Ourahou, Hind, 206

P
Partington, Kate, 44
Partners for Automated
Vehicle Education
(PAVE), 8
Personal occupancy vehicles
(POVs), 91
Personal recognition, 4
Peterson, Lisa, 188
Placemaking, 168
Detroit's neighborhoods
approach and
challenges, 112–113
bankruptcy, 110
funding strategy, 114
goals and objectives,
111–112

infrastructure
improvements, 110
Ford Motor Company
approach and
challenges, 107–108
goals and objectives,
106–107
mobility innovation
and automotive
technology, 108
mobility revolution, 106
principles, 106–107
redevelopment
projects, 109
Light Rail Transit (LRT)
line
approach and
challenges, 116–117
goals and
objectives, 116
locally preferred
alternative
(LPA), 115
transit-oriented
developments
(TODs), 117
TSAAP process, 117
planning profession, 104
urban and regional
planners, 104
PlanetM Mobility Grant
program, funding,
41, 42
Airspace Link, 43
approach and
challenges, 42
Derq, 43
goals and objectives, 42
Pratt Miller, 43
Propelmee, 43
PLUM Catalyst, 191–193
Pluszczynski, Carolina,
109–110
Policy and legislation
automobile development,
manufacturing and
sales, threats to,
120–121
Commission on the
future of mobility

approach and
  challenges, 123–125
goals and
  objectives, 122
workforce
  transition, 122
development patterns
  and transportation
  systems, 120
NEPA, 120
"Remixing Innovation
  for Mobility Justice"
  community needs, 132
  equitable changes to
    bus service, 130
  goals and objectives, 128
  isochrone map, 133
  new park serving local
    low-income
    earners, 129
  oral interviews and
    histories with
    data, 131
  potential service
    adjustments, 133
  public transit-shared
    mobility, 128
  Remix Stats Grab tool,
    132–133
Public recognition, 4

Q
Quigley, Tamy, 53

R
Regional Transit Authority
  (RTA), Southeast
  Michigan Ballot
  Initiative, 6
  approach and
    challenges, 47
  goals and objectives,
    45–47
  mission, 45
  transit millage, 48
  transportation master
    plan, 46, 47
Register, Rajoielle "Raj,"
  87–88
Revolutions per minute
  (RPM) to
  experiences per mile
  (EPM)

approach and challenges,
  36
EPM Advisory Council,
  36–38
goals and objectives, 36
HARMAN and SBD
  Automotive, 35, 36
Rickwalt, Misty, 53
Rooney, Kathleen, 151–152
Roraff, Amanda, 44
Roscini, Julie, 110
Rossi, Maria-Luisa, 171–172
Rotary Club of
  Towsontowne
  approach and challenges,
    200
  goals and objectives,
    199–200
  GoFundMe campaign,
    200
  Navajo Nation, 199
  White Mountain Apache
    community, 199
RTA Rapid Transit Line, 46
Russell, Tammy Meehan,
  191–193

S
Scheinman-Wheeler,
  Nancy, 200–201
Science, technology,
  engineering, and
  math (STEM)
  workforce, 22
SHE-MOVES, 40
Simmons, Ashley, 126
Smith, Courtney, 84
Snorrason, Kathryn, 44
Southeast Michigan
  Council of
  Governments
  (SEMCOG), 48–49
Streets programs, 6
Sustainability and climate
  resilience
  climate preparedness and
    resilience exercise
    series
    approach and
      challenges, 155–156
    flooding, Hurricane
      Harvey, 153

goals and objectives,
  154–155
Houston–Galveston,
  TX, 154
private sector
  organizations and
  nongovernment
  organizations,
  156–157
"scenario planning,"
  158
equity through electrified
  micromobility
  micromobility,
    overregulation
    of, 140
  nonprofit
    organizations
    (NGO), 140
  footprint visibility, lack
    of, 137
  GHG emissions, 138
  gross domestic product
    (GDP), 139
Hawaii, sustainable and
  autonomous
  future in
  approach and
    challenges, 150
  GHG emissions, 148
  goals and objectives,
    149–150
  Honolulu's
    autonomous rail
    transit system, 151
  public electric vehicle
    charging
    station, 149
  Ulupono Initiative, 150
Paris Accord treaty, 138
women and BIPOC, 139

T
Talent and education
  diversity, equity, and
    inclusion (DEI)
    tools
    approach and
      challenges, 164
    CADIA, 163–165
    champion diverse
      talent, 164

Index **233**

drive systemic change, 163–164
initiatives, 163
support leadership commitment, 164
gender disparities, 161
mobile future, designing approach and challenges, 167–169
care exchange scenario, 169
College for Creative Studies (CCS), 166, 171
entry point, 168–169
goals and objectives, 166–167
localized and mobile health-care facilities, 170
placemaking, 168
NEXT Education approach and challenges, 173–174
goals and objectives, 172–173
NEXT/ITE Technician and Installer Credential program, 174
unmanned aerial vehicles (UAV), 172
Society of Women Engineers (SWE), 162
STEM studies, 162
Technology innovation autonomous bus pilot deployment approach and challenges, 190
CAV technology, 188
EZ10 model, 189
G.E.T. Collaborative, 193
goals and objectives, 189

Intelligent Transportation System (ITS), 188
Minnesota Automated Shuttle Bus, 194
MnDOT, 188
pedestrians, vehicle interactions with, 191
shuttle's performance-based findings from 2017–2018 pilot, 190–191
autonomous trucking, 2018 approach and challenges, 195
Driverless Autonomous Trucks, 196
ecosystem, 194
goals and objectives, 194–195
mobility practice, future of, 196
CASE technology, 183, 184
intelligent air mobility advanced air mobility (AAM), 185
approach and challenges, 186–187
FAA approval, 187
goals and objectives, 186
large-scale commercial drone operation, 185
UAVs, 185, 186
Testing Grants, 42
Thompson, Cheryl, 165
Traffic analysis zones (TAZ), 10
Transportation Analysis District (TAD), 10

Transportation Networking Companies (TNC), 91, 92
Tribal Transportation Program (TTP) Planning funds, 50
Tunnel boring machine (TBM), 64

U
Unmanned Aircraft System (UAS) digital infrastructure, 186
Unmanned traffic management (UTM), 187
U.S. Department of Transportation (USDOT), 25

V
Valenti-Amodeo, Vittoria, 96
Vance, Kendee, 53
Varnadore, Hilari, 147
Voytek, Alexa, 57

W
Walker, Katie, 117–118
Wasley, Emily, 157–159
Webb, Kimberly, 65–66
Weinberger, Penelope, 11
Welch, Kristin, 203
Westervelt, Marla, 126
Wheelchair-Accessible Vehicle (WAV), 93
World Resources Institute (WRI), 13

Y
Young, Jovina, 81

Z
Zack, Rachel, 134–135
Zero-emission vehicles (ZEVs), 72